Principles of Statistical Physics and Numerical Modeling

Principles of Statistical Physics and Numerical Modeling

Valeriy A Ryabov
National Research Centre 'Kurchatov Institute', Moscow, Russia

IOP Publishing, Bristol, UK

Multimedia content is available from http://iopscience.iop.org/book/978-0-7503-1341-4.

ISBN 978-0-7503-1341-4 (ebook)
ISBN 978-0-7503-1342-1 (print)
ISBN 978-0-7503-1343-8 (mobi)

DOI 10.1088/978-0-7503-1341-4

Version: 20180801

IOP Expanding Physics
ISSN 2053-2563 (online)
ISSN 2054-7315 (print)

British Library Cataloguing-in-Publication Data: A catalogue record for this book is available from the British Library.

Published by IOP Publishing, wholly owned by The Institute of Physics, London

IOP Publishing, Temple Circus, Temple Way, Bristol, BS1 6HG, UK

US Office: IOP Publishing, Inc., 190 North Independence Mall West, Suite 601, Philadelphia, PA 19106, USA

Contents

Preface

'Computers changed the world'. This statement is fully applicable to statistical physics, which has changed qualitatively over the last three decades due to the widespread use of numerical calculations based on various thermodynamic ensembles. In place of hypotheses and analytically solved models, which are rather difficult from the mathematical point of view, easy calculations in terms of molecular dynamics and Monte-Carlo methods have become available. These calculations allow one to reproduce practically all the results of statistical physics, and indeed, many of them cannot be obtained by means of pen and paper at all.

The accessibility of direct numerical calculations for all kinds of thermodynamic ensembles, e.g. in application to equation-of-state, phase transitions, diffusion and chemical processes, gives even less-advanced students with some programming skills a powerful tool to understand all the key outputs and concepts of statistical mechanics and the nature of different types of ensembles. This accessibility coupled with the comparatively minor effort required is particularly important for students and non-specialists—considering that even many well-educated specialists familiar with standard courses on statistical mechanics retain a vague sense that the field is somehow connected with quantum mechanics and an unimaginably large number of particles $\sim 10^{23}$, neither of which is true. The behavior of an ensemble containing only a few hundred particles already has some features related to ergodic behavior and the absence of the Planck constant in the results would suggest sufficient reason not to use quantum mechanics at all. Nevertheless, the majority of textbooks, including the well-known course of Landau and Lifshitz (1980), starts with a quantum mechanical picture of the density of states (however, fairly believing that the energy space representation, instead of the phase space one, is a more convenient means of description).

Unlike those textbooks, this book allows the reader to look at the world of statistical physics from the point of view of simple models available in numerical realizations, such as, for example, the model of a 1D chain of interacting particles (the model of Fermi–Pasta–Ulam). Familar from school textbooks, the equations of motion of the particles converted to few lines of the Verlet algorithm enable us to simulate their motion. The molecular dynamics (MD) method built on this algorithm is suited to modeling a great number of materials and processes, and many MD interactive programs are free to access and convenient to use. As a result, the complexity of obtaining qualitative results becomes comparable with such 'work-horses' of theoretical physics as the harmonic oscillator.

It should be added that the majority of software products (including FORTRAN) contain a sequence of random numbers generated in a process of dynamics that is *a priori* known. This means that all results of numerical calculations, including those in this book, are precisely reproducible.

Experts will find here a description of the current state of statistical physics based on a systematic use of the microcanonical ensemble (Dieter 2001). Its application allows direct calculation of thermodynamic values (Pearson *et al* 1985), including the entropy. In comparison to traditional methods, microcanonical thermodynamics allows the tracking of the formation of average values in dynamics, with any appropriate level of visibility, thus demonstrating all the key aspects of the traditional approach.

It should be noted that the literature on statistical physics comprises several hundred books written by brilliant scientists. Many of them have contributed significantly to this fundamental branch of theoretical physics. Therefore, it would seem difficult to expect novelty in such a hackneyed textbooks discipline as statistical physics. However, one example of novelty in this book is the definition of microcanonical temperature. In practically all textbooks on statistical mechanics and thermodynamics, the temperature is introduced as the derivative of energy on entropy. However, theoretical investigations over the last few decades, initiated mainly by the results of numerical modeling of ensembles, have shown that the rejection of this definition as the initial one can expand the concept of temperature and allow its use for the direct calculation of entropy and free energy (Rugh 1997).

In this book, the formulation of another basic concept, pressure, also differs from the conventional presentation. It is based on the introduction of additional degrees of freedom for a description of deformation in thermodynamical ensembles. It is also a new product of the development of statistical mechanics of the last few decades. The derived numeric approximation allows one to consider the volume as well as deformation of a more general kind from a single point of view. That is important in view of the rapidly growing number of calculations referring to general kinds of stress. Therefore, in addition to the compression case currently met in traditional treatment of the isobaric ensemble, other kinds of strain such as elongation and shear are also considered. And the notions of stretch ratio and tension are better suited for this purpose than routine volume and pressure.

This book is organized as follows. First, on the basis of a simple physical model, for example, a chain of interacting atoms, we carry out a numerical ‘experiment’. Then the simulation results, typically a time evolution of the ensemble, are explained in terms of microcanonical, canonical or other ensembles.

In this regard, it would be interesting to give vent to imagination and to visualize what one of the founders of thermodynamics, Boltzmann, would make of learning that all his ideas and hypotheses were tested and proven with the help of computer modeling at the turn of the second millennium. Perhaps he would have sat down at the computer and mastered the MD method (see figure 0.1).

Figure 0.1. Boltzmann at a computer. Image from Dellago S and Posch H A 2008 *Realizing Boltzmann's Dream: Computer Simulations in Modern Statistical Mechanics. Boltzmann's Legacy* (Zürich: European Mathematical Society Publishing House). Reproduced courtesy of Dr Bernhard Reischl.

In order to better assimilate the material, the reader is advised to repeat the numerical calculations that illustrate the results of each section and solve the problems at the end of each chapter. Their solutions and details of the calculation algorithms, which do not have a sufficient priority to be presented in the main text (and are therefore likely to be skipped on first reading), are located in the appendices, where they can be accessed via hyperlinks in the text. The appendices also contain the results of calculations concerning ensembles of greater dimensions (2D and 3D), which qualitatively have no differences to the 1D ensemble, but may be of interest to the more experienced reader. In addition, the reader who wishes to further study the issues and topics raised is referred to the literature, with references available at the end of each chapter.

Although one of the main priorities of this book is the simplicity of the presentation of content, all the results of numerical simulations are accompanied by comprehensive theoretical treatment, requiring from the reader a basic knowledge of classical mechanics and mathematics on the graduate level. Therefore, the assumed audience extends, first of all, to students who will find a clear and relatively brief description of the subject, illustrated by numerous figures and animations. The book will also be useful for professionals wishing to receive not only an extremely vivid and contemporary understanding of the key concepts of statistical physics, but also a guide for using these concepts to directly model the properties of materials.

References

Dieter H E 2001 *Microcanonical Thermodynamics. Phase Transitions in 'Small' Systems* (Lecture Notes in Physics vol 66) (Singapore: World Scientific)

Landau L D and Lifshits E M 1980 *Statistical Physics Course of Theoretical Physics* vol 5 3rd edn (Oxford: Pergamon)

Pearson E M, Halicioglu T and Tiller W A 1985 *Phys. Rev.* A **32** 3030

Rugh H H 1997 *Phys. Rev. Lett.* **78** 772

IOP Publishing

Principles of Statistical Physics and Numerical Modeling

Valeriy A Ryabov

Chapter 1

Phase space

The state of a *closed system* consisting of N particles of weight m in an external field and interacting with each other is completely determined by their coordinates x_i and momenta $p_i = m\dot{x}_i$. These values obey the *equation of motion*, which can be written in two ways. One of them, Newton's equation, establishes the link between the acceleration of a particle (the second derivative of the coordinate on time t) with the force F_i acting on it:

$$m\ddot{x}_i = F_i = -\partial U / \partial x_i. \tag{1.1}$$

For 1D motion, the index i runs the values $i = 1 \div N$. In the case of greater dimensions $\kappa = 2$ or $\kappa = 3$, the summation on the index $\alpha = 1 \div \kappa$ is added. Since all components in this case are independent, we will also sometimes use only one index $i = 1 \div N\kappa$. The potential of a particle interacting with an external field and other particles, in principle, may depend on all coordinates: $U = U(\{x_i\})$.

The second order differential equation of (1.1) follows from the *Lagrangian*

$$\mathscr{L} = \frac{1}{2m}\sum_i p_i^2 - U, \tag{1.2}$$

which allows a transformation of Cartesian x_i to any curvilinear coordinates q_i with respect to the generalized momentum $\partial L / \partial \dot{q}_i$. The result of this generalization of (1.1) may be written as

$$\frac{d}{dt}\left(\frac{\partial \mathscr{L}}{\partial \dot{q}_i}\right) = \frac{\partial \mathscr{L}}{\partial q_i}. \tag{1.3}$$

In statistical mechanics, however, there is a more convenient approach based on the Hamiltonian function

doi:10.1088/978-0-7503-1341-4ch1

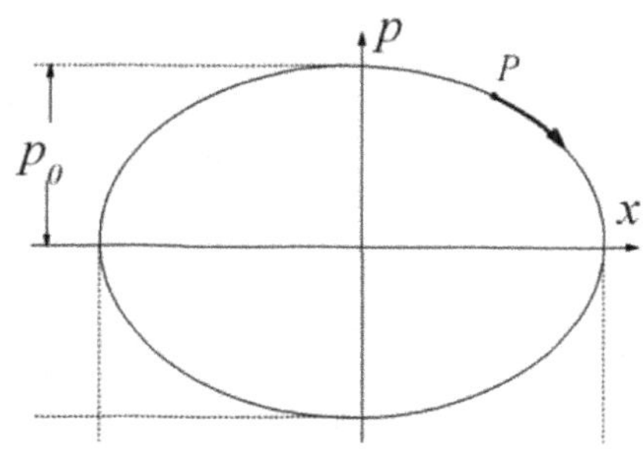

Figure 1.1. The trajectory of the oscillator in phase space.

$$\mathcal{H}(\{x_i, p_i\}) = \frac{1}{2m}\sum_i p_i^2 + U, \tag{1.4}$$

in which the coordinates x_i and impulses of particles p_i are independent variables. A set of them forms the *phase space* of $2\kappa N$ independent variables in which the *state of the ensemble* at any moment of time is given by the $2\kappa N$ vector[1]:

$$(\mathbf{x}, \mathbf{p}) = (x_1, x_2, x_3, \ldots, x_{\kappa N}; p_1, p_2, p_3, \ldots, p_{\kappa N}). \tag{1.5}$$

In terms of this Hamiltonian formalism, the equations of motion of (1.1) can be represented as a system of two differential equations of the first order

$$\dot{x}_i = \frac{\partial \mathcal{H}}{\partial p_i}, \quad \dot{p}_i = -\frac{\partial \mathcal{H}}{\partial x_i}. \tag{1.6}$$

A solution of these equations of motion, being identical to the solution of equation (1.1), gives in the phase plane a curve, the *phase trajectory*, which is presented in the parametrical form as $(x_i(t), p_i(t))$. For example, for a particle in a parabolic potential well $U(x) = m\omega^2x^2/2$ this trajectory is represented by the point $(\tilde{x}, \tilde{p}) = (x_0 \cos \omega t, -\omega x_0 \cos \omega t)$ moving on the ellipse

$$\left(\frac{x}{x_0}\right)^2 + \left(\frac{p}{p_0}\right)^2 = 1 \tag{1.7}$$

with a frequency ω. Here $x_0 = x(0)$ and $p_0 = m\omega x_0$ are the semi-axis (figure 1.1).

Figure 1.2 illustrates the difference between the two approaches. The trajectory at the bottom is the solution of (1.1) and the solution of (1.6) is at the top.

The ellipse form is associated with the energy of a particle preserved during time evolution. For a system of N particles this energy

$$E_N = \frac{1}{2m}\sum_i p_i^2 + U(\{x_i\}) \tag{1.8}$$

[1] Here and below, one neglects their dependence following from conservation of the total and angular momentum.

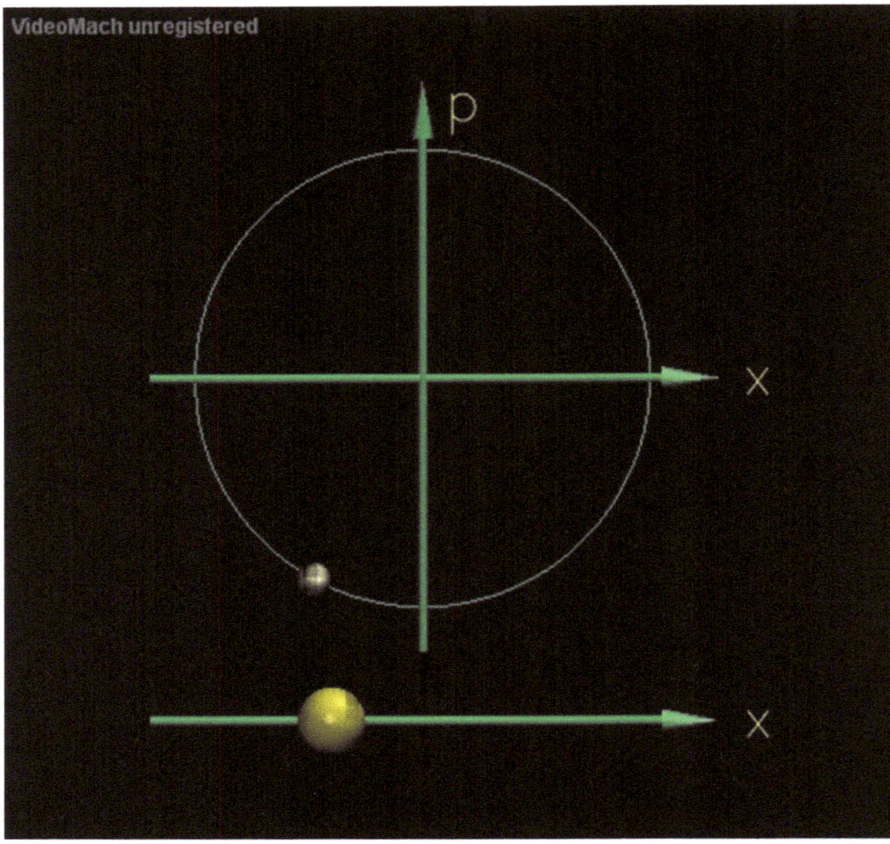

Figure 1.2. Trajectories in terms of the Lagrange and Hamiltonian formalisms. Animation available at http://iopscience.iop.org/book/978-0-7503-1341-4.

formally coincides with the Hamiltonian function. But we need to keep in mind that the equation $\mathscr{H}(\mathbf{x}, \mathbf{p}) = E$ turns to identity only if $\mathbf{x}$ and $\mathbf{p}$ obey the equations of motion.

The set of all trajectories for any initial conditions, the *phase portrait* of the system, gives a visual representation of all the types of evolution and movement. It is particularly important in the case of a system of particles with different energies. In particular, the phase portrait of a system of harmonic oscillators whose energy continuously fills the interval $(E, E + \Delta E)$ is a ring of ellipsoidal form of thickness ΔE (from the point of view of differential geometry, this thickness is determined by the length of a curve, each point of which crosses the ellipse of the fixed energy at a right angle).

Problem 1.1. Construct the phase trajectory of the particle masses $m = 1$ with the energy of (a) $E = 0.5$ and (b) $E = 1.5$ in potential well $U(x) = 1 + \tanh(x)$.

Further reading

Landau L D and Lifshits E M 1969 *Statistical Mechanics A Course of Theoretical Physics 1* (Oxford: Pergamon)

Sussman G J, Wisdom J and Mayer M E *Structure and Interpretation of Classical Mechanics* (Cambridge, MA: MIT Press)

IOP Publishing

Principles of Statistical Physics and Numerical Modeling

Valeriy A Ryabov

Chapter 2

Evolution of the phase space distribution function

Let us introduce the *distribution function* by coordinates and impulses for a system of N particles $f(\mathbf{x}, \mathbf{p}, t)$ (or *phase probability density*) so that value $f(\mathbf{x}, \mathbf{p}, t)\, d\mathbf{x}\, d\mathbf{p}$ is the probability of finding particles in the rectangle $(\mathbf{x}, \mathbf{x} + d\mathbf{x})$, $(\mathbf{p}, \mathbf{p} + d\mathbf{p})$ in time t. As we saw, the trajectory in vector notation $(\tilde{\mathbf{x}}(t), \tilde{\mathbf{p}}(t))$ is defined by the solution of equation (1.6) with initial coordinates $(\mathbf{x}_0, \mathbf{p}_0)$:

$$\dot{\tilde{\mathbf{x}}} = \frac{\partial \mathscr{H}(\tilde{\mathbf{x}}, \tilde{\mathbf{p}})}{\partial \tilde{\mathbf{p}}}, \quad \dot{\tilde{\mathbf{p}}} = -\frac{\partial \mathscr{H}}{\partial \tilde{\mathbf{x}}}. \tag{2.1}$$

Formally, the distribution function can be written as the product of the delta functions

$$\hat{f}(\mathbf{x}, \mathbf{p}, t) = \delta(\mathbf{x} - \tilde{\mathbf{x}}(t))\, \delta(\mathbf{p} - \tilde{\mathbf{p}}(t)). \tag{2.2}$$

It depends on time implicitly, through the time dependence of $(\tilde{\mathbf{x}}(t), \tilde{\mathbf{p}}(t))$. Since the delta function in (2.2) depends on the difference $\mathbf{x} - \tilde{\mathbf{x}}(t)$

$$\frac{\partial \hat{f}}{\partial \tilde{\mathbf{x}}} = -\frac{\partial \hat{f}}{\partial \mathbf{x}}, \tag{2.3}$$

the same holds for the momenta, so

$$\frac{\partial \hat{f}}{\partial t} = \frac{\partial \tilde{\mathbf{x}}}{\partial t}\frac{\partial \hat{f}}{\partial \tilde{\mathbf{x}}} + \frac{\partial \tilde{\mathbf{p}}}{\partial t}\frac{\partial \hat{f}}{\partial \tilde{\mathbf{p}}} = -\frac{\tilde{\mathbf{p}}(t)}{m}\frac{\partial \hat{f}}{\partial \mathbf{x}} + \frac{\partial U(\tilde{\mathbf{x}})}{\partial \tilde{\mathbf{x}}}\frac{\partial \hat{f}}{\partial \mathbf{p}}. \tag{2.4}$$

Finally, shifting the right-hand side of this equation to the left, we obtain the equation for the total derivative on time

$$\frac{d\hat{f}(\mathbf{x}, \mathbf{p}, t)}{dt} = \frac{\partial \hat{f}}{\partial t} + \frac{\tilde{\mathbf{p}}(t)}{m}\frac{\partial \hat{f}}{\partial \mathbf{x}} - \frac{\partial U(\tilde{\mathbf{x}})}{\partial \tilde{\mathbf{x}}}\frac{\partial \hat{f}}{\partial \mathbf{p}} = 0. \tag{2.5}$$

This *Liouville equation* implies the distribution function is constant. Thereby we have proven the *Liouville theorem*, the law of conservation of the density of points in phase space or, in other words, the preservation of the area of the phase space with time. In the mechanics of liquids, the invariance of density during flow signifies incompressibility.

Let us illustrate the Liouville theorem on the example of a 1D ensemble of oscillators, whose initial phase coordinates fill a square of small size (figure 2.1). Each starting point (x_0, p_0) goes to a point (x, p) in time t. This transformation leaves the density of points unchanged. Therefore, despite the distortion of the initial shape, its area remains the same.

The energy conservation for particles with initial vector $\mathbf{x}_0$, $\mathbf{p}_0$ and total energy $E = \mathbf{p}_0^2/2m + U(\mathbf{x}_0)$ requires that all components of the vector $(\tilde{\mathbf{x}}(t), \tilde{\mathbf{p}}(t))$ belong to an *isoenergetic surface*

$$\mathscr{H}(\mathbf{x}, \mathbf{p}) = E. \tag{2.6}$$

However, if in the case of harmonic oscillators the filled area of phase space remains a parallelogram and with the period $T = 2\pi/\omega$ transforms again into a square, in the case of *anharmonic oscillators* the pattern changes.

Figure 2.2 shows the distribution function for a system of $N = 62\ 500$ independent 1D oscillators (details of the calculation are given in appendix A.1). The starting phase points in figure 2.2 also uniformly fill a small square (enlarged in the upper inset). Each point in this figure represents the position of a single particle in the phase space in a specified time. The difference in particle energies $E = p^2/2 + U(x)$ lying in the narrow band $E \sim 0.01 \pm 0.001$ and interacting with an external field $U(x) = x^2/2 + 2x^3/3$ are indicated by color.

After several oscillations, the initial square is deformed at first to a curvilinear quadrangle (shown enlarged in the lower inset) and eventually stretches into a line. This occurs because the period of vibrations of an anharmonic oscillator depends on energy. When it is larger (shown in blue), the period is smaller, and the period increases through the colors to red.

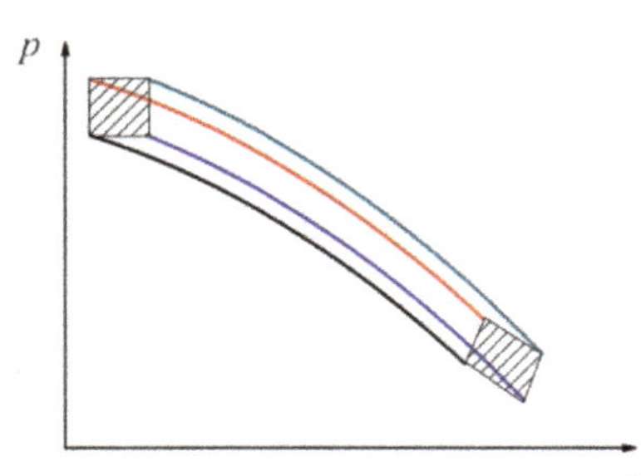

Figure 2.1. Evolution of the phase space element.

After a long time, the distribution ceases to change and spreads across all available areas of phase space (figure 2.3). If you increase each fragment of phase space (one of them is shown at the center of the figure), it is seen that the distribution of points on a phase portrait of the system is practically uniform. The animation in

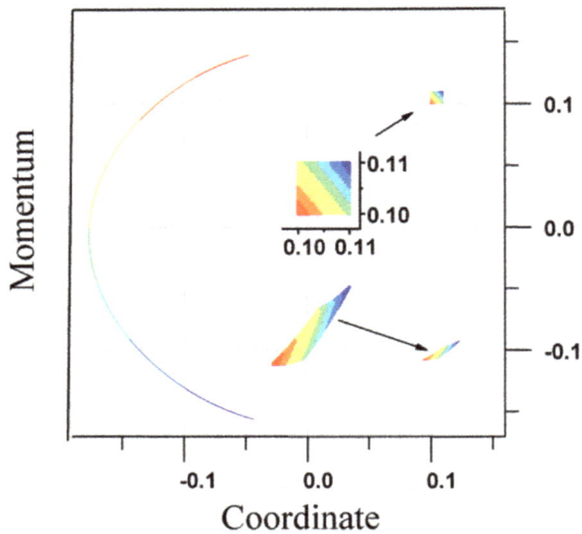

Figure 2.2. Evolution of a system of independent anharmonic oscillators. The initial points of the phase space fill the square (enlarged in the upper inset); deformation of the square after a few oscillations $t = 21$ is shown by a curvilinear quadrangle (enlarged in the lower inset); at $t = 300$ it practically transforms into a curve.

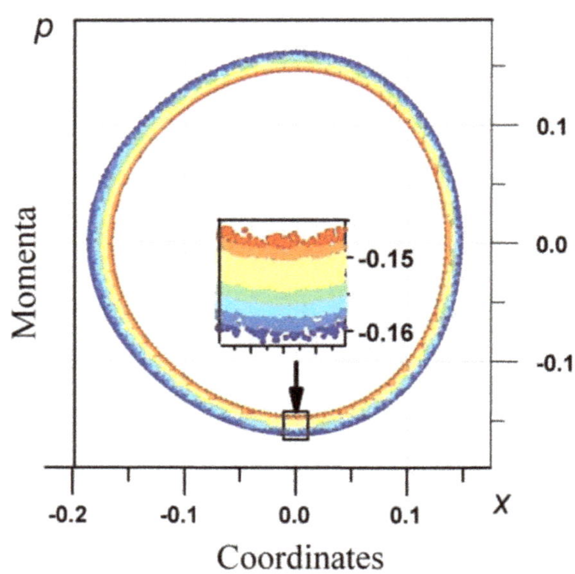

Figure 2.3. Equilibrium distribution: $t = 10^5$.

figure 2.4 illustrates the transition to such a uniform distribution, considering the trajectories of only $N = 10$ particles in phase space. Despite the coordinates of the particles and the distances between them being purely periodic functions (after a certain period of time, they converge almost to a point), the difference in the period after a large number of vibrations results in a large spread of points over the whole phase space. Such a uniform picture should be obtained if the points (x_i, p_i) are 'smeared' in a random way on each closed trajectory. In other words, after a long time, something like a process of chaotization of the *phases of trajectories* (lengths of trajectories on each closed curve) occurs.

In this regard, the behavior of average values is also indicative. In figure 2.5 the evolution of total kinetic and potential energy is shown. Recall that for one oscillator we observe the full pumping from kinetic to potential energy and vice versa. It can be seen now that starting from some period $t > t_{eq} \sim 1000$, the values of K and U become almost constants, coinciding with their averages over any rather long interval of time (see also the comments in appendix A.2).

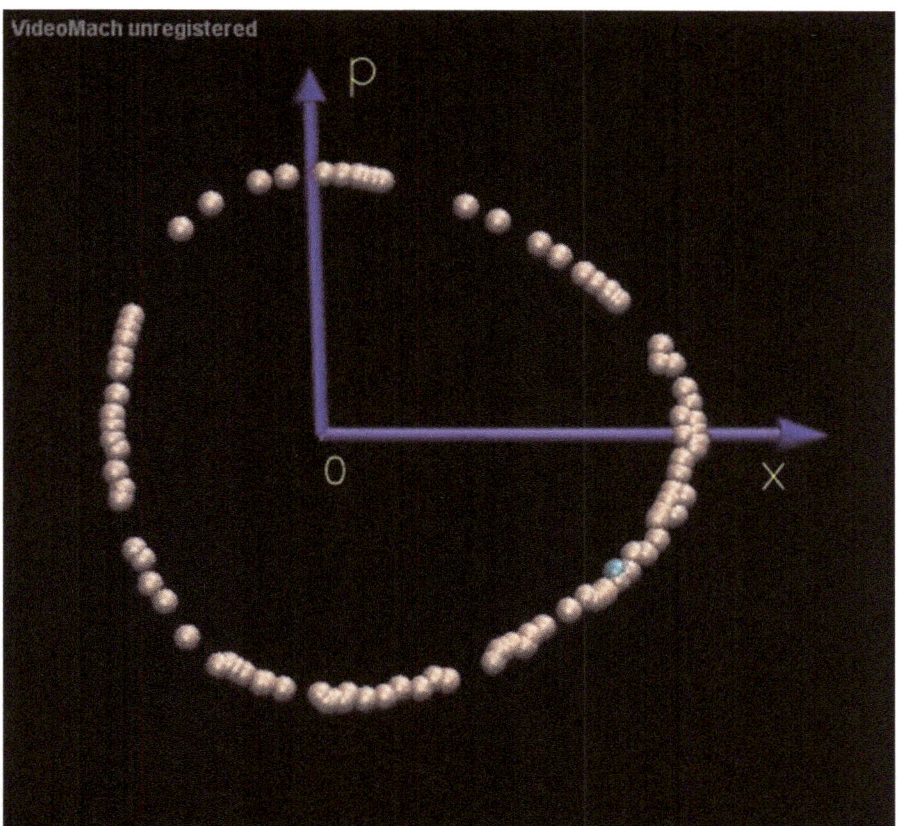

Figure 2.4. Evolution to equilibrium distribution. Animation available at http://iopscience.iop.org/book/978-0-7503-1341-4.

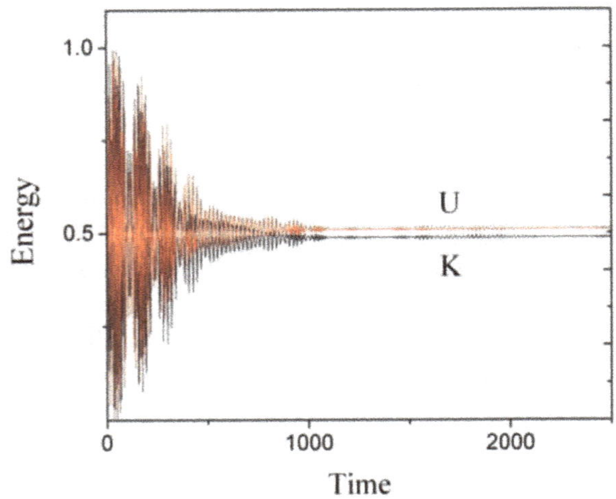

Figure 2.5. The evolution of the kinetic and potential energy of a system of independent 1D anharmonic oscillators in units of the total energy.

The conclusion of the independence of an exact solution at large times of evolution on initial conditions, seen in the behavior of the mean values of energies of the system (figure 2.5), is also confirmed by any variation of initial coordinates on phase trajectories (but preserving the energy of each particle). Our calculations show the independence of the phase space distribution of the time after reaching equilibrium. It follows that the average values tend to limits equal to the average values over the phase space or time. For example, the kinetic energy

$$\overline{K} = \frac{1}{T}\int_0^T K(t)\, dt = \langle K \rangle = \int K(\mathbf{p}) f(\mathbf{x}, \mathbf{p})\, d\mathbf{x}\, d\mathbf{p}. \tag{2.7}$$

In the above 1D example, the time-independent distribution, $t \to \infty$ is established as a result of a certain, although small, still finite width of the initial distribution of energy. It occurs despite the coordinates and velocities of each particle being precisely defined as functions of time. More distinctly, the transition to a time-independent distribution is seen with the increase of dimensionality of space when it is established for a sequence of phase points on a trajectory of only one particle.

Problem 2.1. Write down the condition of conservation of the phase space of the two scattered particles.

Problem 2.2. Find the energy distribution for the particle system shown in figure 2.3 when the starting phase points uniformly fill a small square of area Δ^2 near the point $(x_0, p_0) = (1, 1)$, assuming $m = 1$ and $U(x) = x^2/2$.

Further reading

Chipot C and Pohorille A 2007 *Free Energy Calculations: Theory and Applications in Chemistry and Biology* (Berlin: Springer)

Lindhard J 1965 *Influence of Crystal Lattice on Motion of Energetic Charged Particles Kgl. Dansk. Vid. Selsk.* **34** 14

IOP Publishing

Principles of Statistical Physics and Numerical Modeling

Valeriy A Ryabov

Chapter 3

Equilibrium distributions

The dynamic description of a system of particles in an external potential requires knowledge of trajectories. However, such a precise description as we saw above is redundant. It can be simplified significantly by the assumption that eventually each particle completely 'forgets' its initial coordinate and velocity. Such loss of original information about the system is typical for so-called *ergodic* behavior, when an interaction of any kind provides a stirring of points in phase space. As we have seen in the previous chapter, a similar transition from the dynamic mode to the ergodic mode or chaos is possible even without the participation of any random factors. In the 1D ensemble of independent oscillators considered in chapter 2, the reason for this transition was the anharmonic form of the potential. In this chapter, we will see how an ergodic transition happens for a single particle during its movement in the 2D periodic potential $U(\mathbf{r})$ shown in figure 3.1. For proof, calculate the spatial distribution of one particle in a lattice over a sufficiently long period of time. However, before we do this, let us try to understand what kind of phase distribution is expected in the ergodic mode.

Randomization of the phases of trajectories resulting in distribution function f, independent of initial phase coordinates, is equivalent to the random choice of the vector $(\mathbf{x}, \mathbf{p})$ at each moment of time. The corresponding distribution function can be derived from equation (2.5) if one neglects the explicit dependence on time:

$$\frac{df}{dt} = \frac{\mathbf{p}}{m}\frac{\partial f}{\partial \mathbf{x}} - \frac{\partial U}{\partial \mathbf{x}}\frac{\partial f}{\partial \mathbf{p}} = 0. \tag{3.1}$$

The solution of this equation is an arbitrary function of the Hamiltonian

$$f(\mathbf{x}, \mathbf{p}) = w(\mathscr{H}(\mathbf{x}, \mathbf{p})). \tag{3.2}$$

To make certain, substitute it into equation (3.1), taking into account equation (1.4).

doi:10.1088/978-0-7503-1341-4ch3

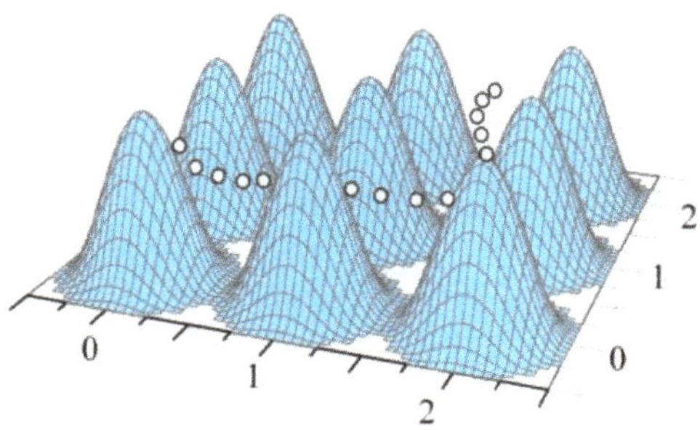

Figure 3.1. The trajectory of a particle in a 2D lattice. Showing the potentials of nine atoms. The height of the path is equal to the energy of the particle. The distances between the points on the trajectory correspond to equal time intervals.

Since all phase coordinates are on the isoenergetic surface $\mathcal{H}(\mathbf{r}, \mathbf{p}) = E$, w is a delta function[1]

$$f(\mathbf{x}, \mathbf{p}) = \delta(E - \mathcal{H}(\mathbf{x}, \mathbf{p})). \tag{3.3}$$

Known as a microcanonical distribution, this distribution generates a microcanonical ensemble.

The set of all trajectories that make up the isoenergetic surface $E = \mathcal{H}(\mathbf{x}, \mathbf{p})$, represents a *phase portrait* of the system. Thus the *i*th particle and, equivalently, each starting point of the square represented in figure 2.3 corresponds to the closed curve $E_i = \mathcal{H}(x_i, p_i)$. This does not mean, however, that the deterministic distribution $\hat{f}(\mathbf{x}, \mathbf{p}, t)$ turns after a long time into the microcanonical distribution $\delta(E_i - \mathcal{H}(x_i, p_i))$. Equilibrium, generally speaking, is not achieved since the distribution function still depends on time. However, shifting each starting point of the square in time along the trajectory, it is easy to see that the visible picture for the whole ensemble does not change substantially. In other words, starting from any coordinates $\mathbf{x}_0$, $\mathbf{p}_0$ on an isoenergetic surface $E = \mathcal{H}(\mathbf{x}_0, \mathbf{p}_0)$ (but leaving the initial distribution of energy intact), we arrive at distribution similar to equation (3.3), being independent of neither the initial coordinates $(\mathbf{x}_0, \mathbf{p}_0)$ nor of time:

$$f(\mathbf{x}, \mathbf{p}) = \int d\mathbf{x}_0 d\mathbf{p}_0 \delta\big(\mathcal{H}(\mathbf{x}, \mathbf{p}) - \mathcal{H}(\mathbf{x}_0, \mathbf{p}_0)\big). \tag{3.4}$$

Let us return to a trajectory of one particle, part of which was obtained by the numerical solution of equation (2.1) with the help of the Verlet algorithm (see appendix A.1), as is shown in figure 3.1. In continuing, however, it will be more convenient to consider the whole trajectory within one cell of the lattice (a so-called *Wigner–Seytz cell*) attached to one atom (shown in figure 3.1). For this purpose, we introduce the *periodic boundary conditions* (PBCs) on all four sides of the square per

[1] Strictly speaking, the well-known result of analytical mechanics suggests the dependence of this function not only on the Hamiltonian but also on other integrals of motion. But here and below, the total momentum and angular momentum are assumed equal to zero, and the change of the number of degrees of freedom associated with these integrals of motion is also neglected.

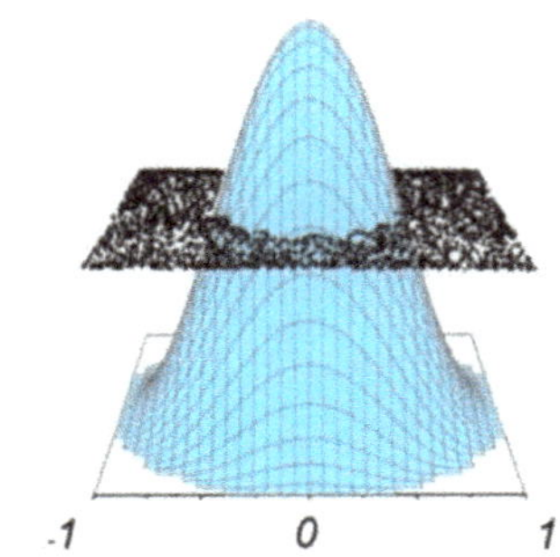

Figure 3.2. The spatial distribution of the particles in periodic potential at long times. All points on the trajectory are placed in a single cell at the height of the particle energy

cell. The PBCs assume the transfer of a particle on reaching the cell border (for example, points $x = 1$, $|y| < 1$) to an equivalent point on the other side of the cell: $x = -1$, y with the same value of velocity (for more detail about PBCs, see appendix A.1). For simplicity, the potential profile is chosen here to be dependent only on the distance to its center r. In addition, it has a zero derivative at points where $U(\mathbf{r})$ vanishes (to avoid a discontinuity in the force while calculating the trajectory).

To visualize the spatial distribution, we will record the positions of the points $\mathbf{r}_i = \mathbf{r}(t_i)$ for times t_i $i = 1 \div I$, separated by a small constant interval (as shown in figure 3.1). From the formal point of view, the space distribution received in this way is equivalent to the procedure of the time average of equation (2.7):

$$\widehat{n}(\mathbf{r}, t) = \frac{1}{t}\int_0^t dt' \, \delta(\mathbf{r} - \tilde{\mathbf{r}}(t')). \tag{3.5}$$

For a small interval of time, the density of points is strongly non-uniform and blank regions of space occur (see figure A.2 in appendix A.2). But after a long time (figure 3.2, right) all recorded points evenly (i.e. with a uniform density $\widehat{n}(\mathbf{r}) = \text{const}$) fill the entire available area of the cell (the latter follows from the equation $E \geqslant U(\mathbf{r})$, where $E = p_0^2/2m + U(\mathbf{r}_0)$ is the energy of a particle). This conclusion can be checked by counting the number of points in each small region of the cell shown in figure 3.2.

Obviously, an ensemble of particles whose initial coordinates $(\mathbf{r}'_0, \mathbf{p}'_0)$ belong to a trajectory $\tilde{\mathbf{r}}(t)$, $\tilde{\mathbf{p}}(t)$ generates the same constant spatial distribution. But a stronger statement is also true. The constant density obtained in calculations means the independence of the distribution $f(\mathbf{r}, \mathbf{p})$ of any initial conditions corresponding to the given energy E. In other words, the transformation of dynamic distribution function $\hat{f}(\mathbf{r}, \mathbf{p}, t)$ into some universal one $f(\mathbf{r}, \mathbf{p})$ is obviously due to the loss of 'memory' about the initial coordinates. As we have already seen above, such a property has a microcanonical distribution of equation (3.3)

$$f(\mathbf{r}, \mathbf{p}) = A \, \delta(E - \mathbf{p}^2/2m - U(\mathbf{r})), \tag{3.6}$$

where A is the normalization constant.

The steady-state (equilibrium) distribution of particles in phase space leads to the spatial distribution

$$n(\mathbf{r}) = \int d\mathbf{p} f(\mathbf{r}, \mathbf{p}) = 2\pi A \int dp\, p\,\delta(E - p^2/2m - U(\mathbf{r})), \tag{3.7}$$

which, after integration, goes to a constant in the accessible area seen in figure 3.1:

$$n(\mathbf{r}) = \frac{1}{S(E)}\vartheta(E - U(r)). \tag{3.8}$$

Here $\vartheta(x)$ is the step function

$$\vartheta(x) = \begin{cases} 1, & x \geqslant 0; \\ 0, & x < 0, \end{cases} \tag{3.9}$$

and $S(E)$ is the accessible area. It is equal to the square of a cell in figure 3.1 after deducting the area of a circle of radius r^*, such that $E = U(r^*)$.

Note that the uniform distribution of equation (3.8) is actually the result of the dimensionality of the coordinate system. The motion in a 1D potential $U(x)$ is not uniform (see problem 3.1).

How do the ergodic transition and, consequently, the transformation of the distribution $\widehat{f} \rightarrow f$ arise? For the example considered above, this question could be paraphrased: how does the small initial area of the phase space extend on all of the *energy surface* $\mathscr{H}(\mathbf{r}, \mathbf{p}) = E$?

This process can be treated on the example of two trajectories which start from two very close points of phase space separated by the distance $\delta_0 = 0.0001$ (one of them is the starting point of a trajectory in figure 3.1, see appendix A.2). Over time they diverge exponentially with time, $\delta_0 = \delta_0 \exp(\lambda t)$ (*Lyapunov's exponent*, see figure 3.3), providing thereby intensive mixing in the phase space and creating

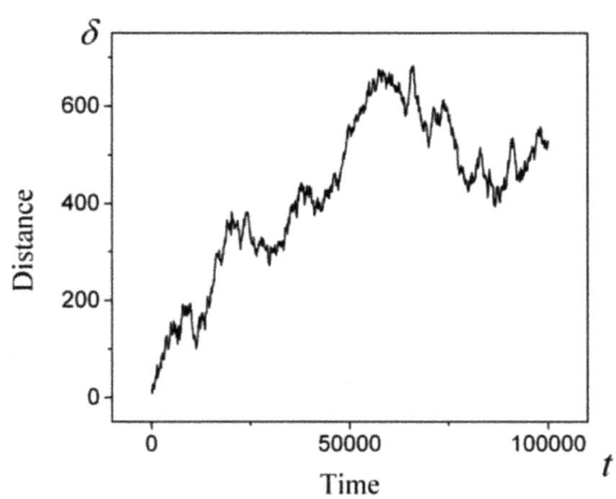

Figure 3.3. Evolution of the distance between two points in phase space.

conditions for the independence of the final distribution of time. Extensive literature (see, for example, Hinchin 2003 and Ryuel 1989) is devoted to this problem.

For our purposes, however, only one circumstance related to this section is essential. The behavior of interacting of particles in an ensemble transforms from a purely dynamic to an *ergodic* one, at which point the system forgets its phase coordinates[2]. This determined chaos appears due to the high sensitivity to small perturbations of the initial conditions. In the 1D case, as we have seen, an anharmonic form of the potential was a key factor for the phase coordinate mixing (in a harmonic approximation, mixing of any kind does not occur). Eventually, the exact time-dependent distribution may be approximated by the microcanonical distribution with very good accuracy. Such an approach essentially simplifies the procedure of finding mean values as, instead of the average over the vectors $(\tilde{\mathbf{x}}, \tilde{\mathbf{p}})$, e.g. over time, it is possible to use a set of random vectors obeying the microcanonical distribution.

A few words should be said about the correspondence between the system of N particles considered hereafter and a continuous phase space system, to which only the notion of 'volume' seemed to make sense. It is assumed that an increase in the number of points in an ensemble accompanied by reduction of the dispersion of average values, does not lead to any new results. Therefore, with a good accuracy, a finite ensemble describes the behavior of a continuous distribution of probability in phase space.

Problem 3.1. Calculate the phase space distribution for a 1D microcanonical ensemble with the initial phase space shown in figure 2.3. Compare the results of both distributions.

Problem 3.2. Find the spatial distribution for the conditions of the previous problem.

Further reading

Chandler D 1987 *Introduction to Modern Statistical Mechanics* (New York: Oxford University Press)

Hinchin Ya A 2003 *Mathematical Bases of Statistical Mechanics* (Moscow: Research Center Regulyarnaya i Haoticheskaya Dinamika)

Reif F 2009 *Fundamentals of Statistical and Thermal Physics* (Long Grove, IL: Waveland)

Ryuel D 1989 *Statistical Mechanics. Rigorous Results* (Singapore: World Scientific)

[2] In the 1D case, anharmonicity is usually sufficient for the mechanism of phase mixing (such mixing does not occur in a harmonic potential).

IOP Publishing

Principles of Statistical Physics and Numerical Modeling

Valeriy A Ryabov

Chapter 4

Ensemble of free particles

The main conclusion of the previous chapter for independent particles in an external field can be extended to the case of an ensemble of interacting particles. As above, we shall study the transition from the dynamic description in terms of the distribution function $\hat{f}(\mathbf{x}, \mathbf{p}, t)$, built on exact trajectories, to the microcanonical distribution of equation (3.6) upon the occurrence of the ergodic mode (e.g. irrelevance to initial conditions).

Consider the collision dynamics of the simplest ensemble particles—a linear chain of particles with periodic boundary conditions. Introduction of periodic boundary conditions allows us to consider our ensemble in the form of a closed system with a finite number of particles. It can be imagined as a chain of atoms bonded by springs so that the length of the arcs connecting atoms plays the role of the distances between the particles (figure 4.1). The length of a circle is equal to Nd, where N is the number of atoms and d is the period of a chain. The origin of the coordinate frame can be chosen at any point of a circle, in figure 4.1 it is marked by the vertical line segment. Despite the 2D geometry of this picture, the dynamics of particles here is defined by only the 1D coordinates of arc length[1].

In considering the dynamics of a many-particle system, it is natural to begin with the simplest case, a gas of free particles, or an ideal gas. But we cannot fully deny interaction or assume there is no dynamics at all. Nevertheless, in statistical mechanics, an ensemble possessing all properties inherent in a gas of free particles can be introduced if we assume the presence of collisions, instant interactions leading to momentum transfer between colliding particles. A uniform spatial distribution will be a criterion of 'freedom' in this case, as it should be for the free particles.

However, the transition from the 1D ensemble of interacting particles represented in figure 4.1 to a special case of an ensemble of non-interacting particles seems non-

[1] Since the bending radius of the chain preserves the arc lengths, they can be regarded not as generalized, but as Cartesian coordinates.

doi:10.1088/978-0-7503-1341-4ch4

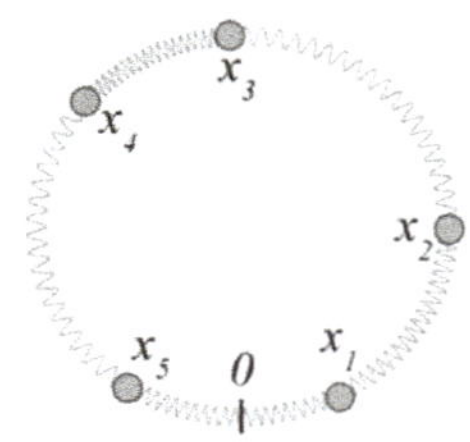

Figure 4.1. On the concept of periodic boundary conditions.

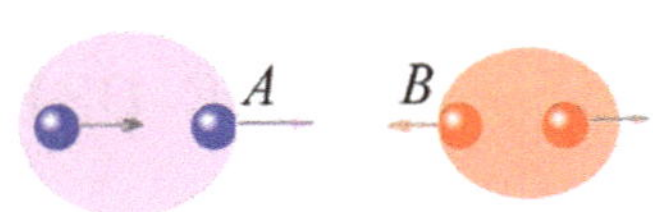

Figure 4.2. Collision of free particles (a pair of atoms) in the chain.

trivial. The model of the chain of atoms shown in figure 4.1, but without springs, with pairs of facing atoms, is not suitable. Head-on collisions of neighboring atoms result only in an exchange of impulses between them, without a change of momentum distribution. It is possible, however, to complicate the collision process, suggesting that in addition to directly colliding particles A and B at one point, their next neighbors are also involved in the collision process by exchanging their impulses (figure 4.2) so that the total energy and momentum of four colliding particles is conserved. For example, assuming a collision occurs at the point $x_2 = x_3$ (see figure 4.2), the neighboring first and the fourth particles also take part in the exchange of impulses. The more detailed diagram of such 'quasimolecular' collisions in figure 4.3 is described in appendix A.2. As in the system of the center of mass, the velocities of the particles after the collision are not necessarily opposite (as occurs in pair collisions), the arrangement of particles can change, making possible the mixing of their coordinates.

Numerical simulations were performed (see appendix A.2) for 1D of the ensemble of $N = 1000$ particles with initial positions in the chain cells $x_i^0 = id$. Initial impulses were sampled according to the uniform distribution in the interval $(-\delta, \delta)$ with $\delta = 0.5$. The average impulse $\bar{p} = \sum_{i=1}^{N} p_i / N$ was afterwards subtracted from p_i, preserving zero momentum of the system.

Figure 4.4 shows the evolution of the spatial distribution. For greater visibility, all atoms are distributed in ascending order of initial coordinates into ten groups marked by a color containing one hundred atoms each. The Y-axis is also introduced, indicating the ordinal position in the group for each atom, so the process of spatial mixing is seen in two dimensions. It is seen that at times $t \sim 3 \cdot 10^5$, the order of atoms in the chain is completely broken and the spatial distribution becomes uniform within the entire length of the chain.

Now turn to the momentum distribution established in these calculations. Its evolution for $0 < t < 30$, when starting from an initial uniform distribution, can be traced back in figure 4.5.

Figure 4.3. Scattering of molecules in a 1D ideal gas. Animation available at http://iopscience.iop.org/book/978-0-7503-1341-4.

One can see how, even long before the mixing of the coordinates, the momentum distribution from the uniform (blue dotted line in figure 4.6) becomes normal with the dispersion $\overline{p^2} = \delta^2/3$ (red curve). So, eventually, the distribution in the phase space becomes

$$f(\boldsymbol{x}, \boldsymbol{p}) = \frac{1}{Nd\sqrt{2\pi\overline{p^2}}} \exp\left(-\frac{p^2}{2\overline{p^2}}\right). \tag{4.1}$$

Note that the average energy per one particle $\overline{p^2}/2 = E/N(m = 1)$ depends on the magnitude of the variance associated with the initial uniform momentum distribution: $\overline{p^2} = \overline{p_i^{02}} = \delta^2/3$. Once again we come to the conclusion that after some time the system forgets the initial conditions and becomes ergodic. In this mode, it obeys some universal distribution of the phase coordinates, independent of time. Since it relates to the system, preserving the total energy, this distribution must be at least a consequence of the microcanonical distribution

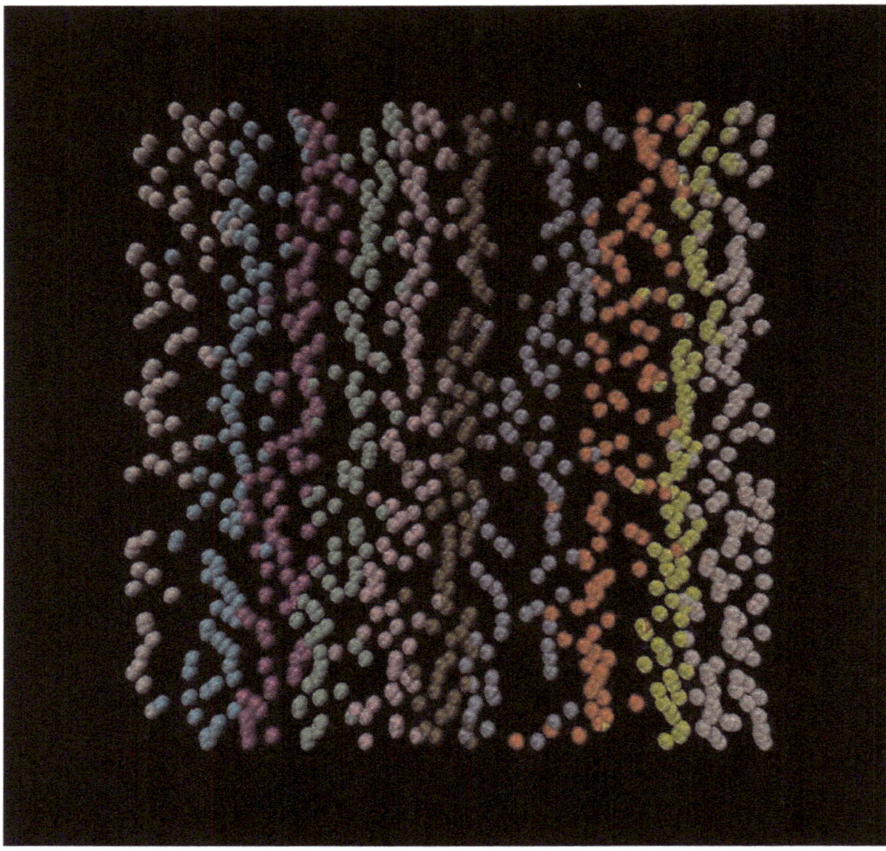

Figure 4.4. Evolution of the spatial distribution of atoms of an ideal gas. Animation available at http://iopscience.iop.org/book/978-0-7503-1341-4.

$$f(\mathbf{r}, \mathbf{p}) = \frac{1}{Nd\Omega}\delta\left(\frac{\mathbf{p}^2}{2m} - E\right), \tag{4.2}$$

where $\mathbf{r}, \mathbf{p} = (x_1, x_2, \dots x_{N-1}, x_N; p_1, p_2, \dots p_{N-1}, p_N)$, and Ω is a normalization constant.

The distribution of equation (4.2), however, indicates only the law of the conservation of energy and has nothing to do with the momentum distribution shown in figure 4.6.

It is also easy to check that the simulation results do not depend on the number of particles in the ensemble, as long as this number is large enough. As can be seen from figure 4.6 and seems intuitively obvious (and could be exactly confirmed by analyzing the obtained results), this distribution is related to any sufficiently large piece of the whole ensemble. Moreover, an even stronger statement still stands, referring not to the distribution of the phase coordinates in the ensemble for a given time, but relating to each particle separately. To verify this it is sufficient to average

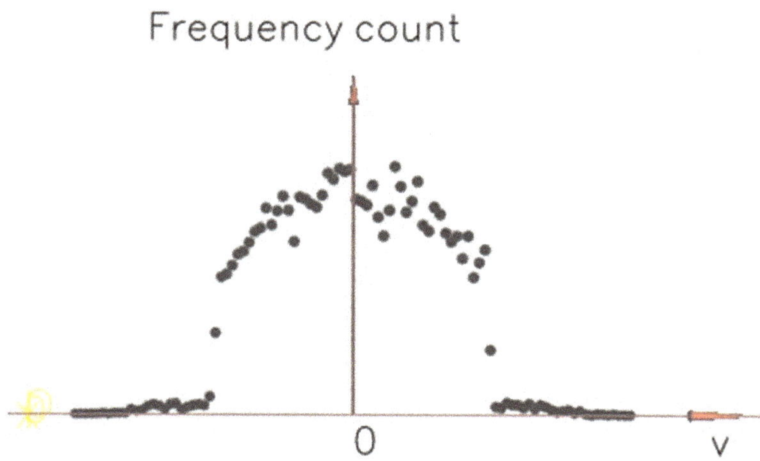

Figure 4.5. Scattering of molecules in a 1D ideal gas. Animation available at http://iopscience.iop.org/book/978-0-7503-1341-4.

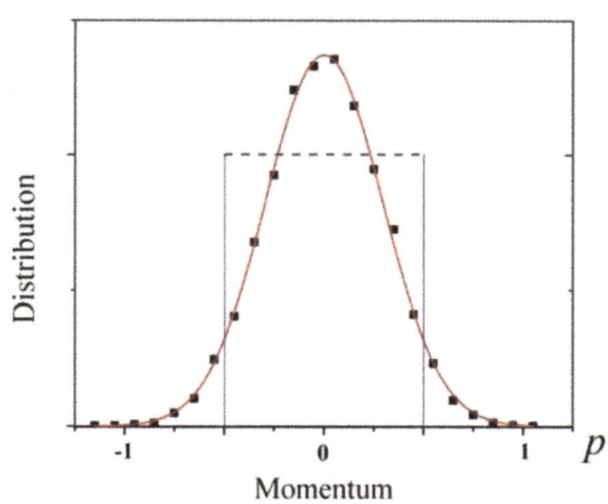

Figure 4.6. Momentum distribution of a gas of free particles at $t = 10^4$. The dashed line is the initial distribution.

the single particle distribution over a large period of time for $t > t_{eq}$ when complete mixing of the phase coordinates takes place. We suggest the reader should extend the corresponding calculations for the times $t \gg 10^5$ and make sure that the distribution of equation (4.1) for each particle goes to the distribution for the whole ensemble as discussed above.

Further reading

Landau L D and Lifshits E M 1980 *Statistical Physics Course of Theoretical Physics* vol 5 3rd edn (Oxford: Pergamon)

ter Haar D 1995 *Elements of Statistical Mechanics* (Oxford: Butterworth-Heinemann)

IOP Publishing

Valeriy A Ryabov

Chapter 5

Ensemble of interacting particles

Consider now the 1D chain of N interacting atoms shown in figure 4.1. The length of the arcs between atoms plays the role of distances between particles, so the length of the chain Nd is equal to the circumference. The equation of motion of (1.6) for the ith atom of the chain becomes

$$p_i = m\dot{x}_i, \quad \dot{p}_i = -\frac{\partial U(x_i)}{\partial x_i}, \tag{5.1}$$

where the interaction potential i– th atom with the $(i - 1)$th and $(i + 1)$th nearest neighbors is given by

$$U(x_i) = U(x_i - x_{i-1}) + U(x_{i+1} - x_i), \tag{5.2}$$

where $U(x)$ is the pairwise potential of the interaction of two atoms at distance x from each other.

The numerical solution of equations (5.1) was made for $N = 2000$ atoms in a chain whose initial velocity is assumed to be zero and mass $m = 1$. The potential $U(x) = \Delta^2/2 - 2\Delta^3/3$, where $\Delta = x_i - x_{i-1} - 1$ was used. The displacement from equilibrium positions $x_i^0 = i\,d$ were simulated due to the uniform distribution in the interval $(-\delta, \delta)$ with $\delta = 0.1$ (see appendix A.3). Such a choice of initial conditions provides the immovability of the chain as a whole since its initial impulse is equal to zero. In addition, the coordinates of moving atoms remain close to their equilibrium positions.

To visualize the phase density, as was done in the previous chapter, fix the points $(x_i^\alpha, p_i^\alpha) = (x_i(t_\alpha), p_i(t_\alpha))$ in the phase plane at the equidistant times $t_\alpha : \alpha = 1 \div I$ (cf with equation (3.5) and figure 3.2). The result for $t = 5000$ is shown in figure 5.1. It is seen that the coordinate of each vibrating ith particle is concentrated near its equilibrium positions x_i^0 with the same spread of an order δ^2, thus making the energies of the particles of the order of $\sim E/N \sim \delta^2$.

doi:10.1088/978-0-7503-1341-4ch5

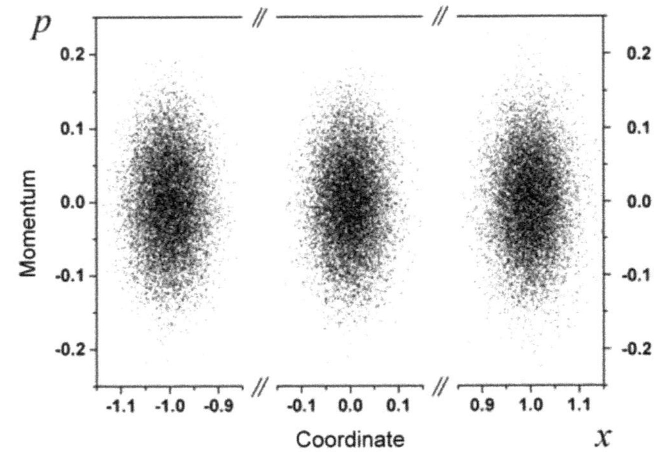

Figure 5.1. The phase portrait of the three neighboring particles in the chain.

Unlike free particles, the spatial distribution here is not uniform, although the distribution of the phase coordinates is the same within each periodic cell. Again, as above, the resulting distribution does not depend on the number of particles in the ensemble and relates not only to the whole ensemble but to any part of it.

Moreover, the distribution in phase space is identical for each particle, being independent of initial conditions. That means that the equilibrium distribution of phase coordinates, for free particles at least, is governed by the microcanonical distribution

$$f(\mathbf{x}, \mathbf{p}) = \frac{1}{\Omega}\,\delta(\mathscr{H}(\mathbf{x}, \mathbf{p}) - E). \tag{5.3}$$

However, the understanding of this fact does not supply us with the answer to the question: what parameters characterize the equilibrium distribution of coordinates and impulses seen in figure 5.1 (similar to figure 2.2)?

Let us turn to the momentum distribution shown in figure 4.6. The frequency counts (see problem 5.1) show that impulses are distributed in accordance with the normal distribution with the variance

$$\overline{p^2} = 2mE/N. \tag{5.4}$$

Consequentially, the value of the kinetic energy per atom also is equal $\overline{K_i} = \overline{p^2}/2m$, which is not surprising if one takes into account the circular symmetry of the arrangement of particles in the chain.

Calculation of the average of the product of any pair of impulses gives $\overline{p_i p_j} \sim 1/\sqrt{N}$. This means that correlations between impulses are absent and, therefore, the distribution of each particle is determined by the same mean square $\overline{p_i^2} = \overline{p^2}$:

$$f(p_i) = \frac{1}{\sqrt{2\pi \overline{p^2}}} \exp\left(-\frac{p_i^2}{2\overline{p^2}}\right). \tag{5.5}$$

This type of distribution shows that it does not depend on the mass of particles, therefore, the average kinetic energy per degree of freedom is equal to $\overline{K_i} = \overline{p^2}/2m$ (the equipartition rule). The independence of momenta from each other means that their joint distribution can be presented by the product of the separated distributions of single particles

$$f_N(p_1, p_2, ..., p_N) = \frac{1}{(2\pi\overline{p^2})^{N/2}} \exp\left(-\frac{\sum_{i=1}^{N} p_i^2}{2\overline{p^2}}\right). \tag{5.6}$$

Thus again, the momentum distribution of each particle, even in the presence of interaction, is given by the same distribution as that of free particles of equation (4.1). But this conclusion does not stand for the coordinates. Calculations show that the distribution function of coordinates seen in figure 5.1 is also close to the normal one with the mean square displacement identical for all particles $\overline{(x_i - x_i^0)^2} = \overline{\Delta x^2}$. However, this does not mean that the joint distribution of the coordinates could be represented as a product of a one-particle distribution, as is the case for momentum distribution. The phase coordinates in the system are *a priori* not independent and obey a more complicated form of joint distribution. In fact, a similar direct check shows that variables x_i are correlated because the correlation coefficients $\overline{x_i x_j} \sim \text{const}\, \overline{x_i^2}$ do not tend to zero. The reason for such behavior of coordinates and impulses will be discussed later in chapters 9 and 10.

Problem 5.1. Find the mean kinetic energy of the particles for a 3D ensemble.

Further reading

Kittel C 1953 *Introduction to Solid State Physics* (New York: Wiley)

IOP Publishing

Principles of Statistical Physics and Numerical Modeling

Valeriy A Ryabov

Chapter 6

Temperature

Emphasize the key conclusion following from the obtained simulation results. As soon as an equilibrium is reached, all dynamic variables and their momentum, being the exact solution of the equation of motion, can be replaced by random numbers with a corresponding distribution, so that each time may be considered as the beginning of their independent generation. In other words, the set of points in phase space of the system, the *microstate*, the population of which is shown in figure 5.1, might be generated according to a time-independent distribution. From this follows the equivalence of averages in time $T \gg 1$ and in phase space of any value $a(t)$ being implicitly time-dependent on coordinates and impulses[1]

$$\bar{a} = \frac{1}{T}\int_0^T a(t)\, dt = \langle a \rangle = \int a(\mathbf{x}, \mathbf{p}) f(\mathbf{x}, \mathbf{p})\, d\mathbf{x}\, d\mathbf{p}. \tag{6.1}$$

This underlines the important conclusion of previous chapters: the momentum distribution for the free and interacting particles is the same. This means some property of universality of the macroscopic average relating to this distribution. Keeping this in mind we introduce for the average kinetic energy per one degree of freedom the quantity T, called *absolute temperature*,

$$kT = 2\overline{K}_i = \frac{\overline{p_i^2}}{m} = \overline{\frac{1}{N}\sum_{i=1}^{N}\frac{p_i^2}{m}}, \tag{6.2}$$

[1] Strictly speaking, the implementation of this relationship appears as a result of the ergodic hypothesis, which suggests that during its evolution the system passes arbitrarily close to any point of the phase space on the constant energy surface, thereby implementing a realization of the full group of events in terms of probability theory. However, as finite systems are the subject of this book, *a priori*, do not assume the filling of phase space in that way.

doi:10.1088/978-0-7503-1341-4ch6

where $k = 1.38 \cdot 10^{-23}$ J K^{-1} is the *Boltzmann constant*, and the value of T is measured in units of kelvin, K. The value under the sign of an average in equation (6.2), the *microcanonical temperature*, is proportional to the instant value of total kinetic energy. Thus, for a given temperature, the joint momentum distribution can be rewritten in the form of the *Maxwell distribution*

$$f_p(p_1, p_2, ..., p_N) = \frac{1}{(2\pi kT/m)^{N/2}} \exp\left(-\frac{\sum_{i=1}^{N} p_i^2}{2mkT} \right). \tag{6.3}$$

In the thermodynamic limit $N \to \infty$ the total energy of an ideal 1D gas is

$$E = \frac{N}{2}kT. \tag{6.4}$$

Let us address now a system of interacting particles. The difference in the behavior of the ensemble at different temperatures is demonstrated by the animation in figure 6.1, which shows the results of the simulation of a 2D hexagonal lattice containing $N = 200$ atoms for two different temperatures. The initial uniform distribution of momenta for each particle was limited by the values $|\delta| = 0.02$ (left) and $|\delta| = 0.2$ (right) (see appendix A.4). You can see the difference in the amplitude and velocity of the atoms.

Talking about the dynamics of a system of greater dimension, one recalls that all degrees of freedom are equivalent because they participate in the equations of motion (2.1) independently. This was the justification for referring with the index i in all formulas above not only to the number of a particle, but also to the index of the spatial coordinate. This generalization on a 3D case is carried out by the replacement $N \to 3N$. For a 3D ensemble, correspondingly, the equipartition rule goes to $\overline{K_i} = E/N = 3kT/2$. In a microcanonical ensemble with interaction, the distinction in the redistribution of the total energy to kinetic and potential energy affects its final temperature.

Figure 6.2 shows the kinetic energy (while conserving the total energy $E = 0.002\,92$) for the atomic chain described in the previous chapter. It can be seen that in some time t_{eq} when equilibrium is reached, the kinetic and potential energies tend to a constant value. Minor fluctuations of the kinetic energy seen in the figure for $t > t_{\text{eq}}$ are the consequence of having a finite number of particles in the chain. The increase in the number of particles by two orders in the ergodic mode of the movement should reduce the amplitude of fluctuations on the order. Such behavior is actually observed in 2D dynamics and 3D lattices[2].

[2] The ergodic mode does not occur in a 1D chain as per the so-called Fermi–Pasta–Ulam problem (appendix A.4). However, purely visually, the calculation results shown in figures 6.1 and 6.2 are very similar to those if the trend towards equilibrium has occurred.

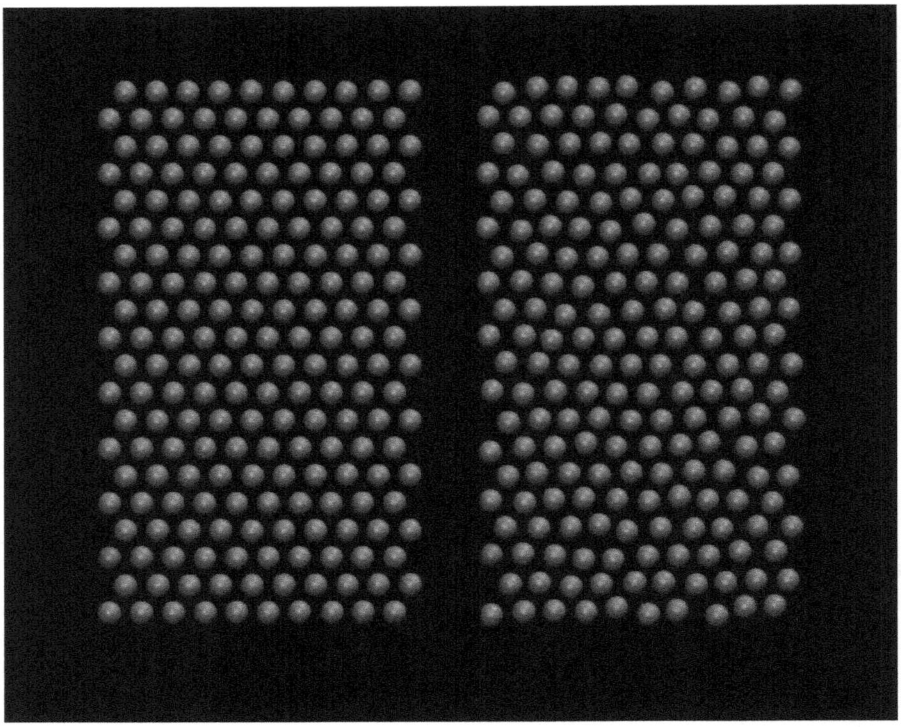

Figure 6.1. Lattice for different temperatures. Animation available at http://iopscience.iop.org/book/978-0-7503-1341-4.

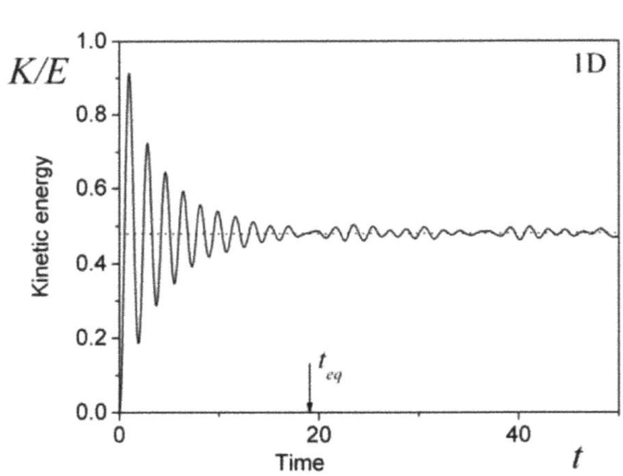

Figure 6.2. The evolution of the kinetic energy of the oscillating chain of particles.

This circumstance is common to all macroscopic quantities. As long as each of them is determined by the microstates being a set of random numbers, any estimate of macroscopic average, according to the central limit theorem, has a variance of an order $\sim 1\sqrt{N}$ regardless of the type of distribution. In addition, since on average each particle should not be different from others, the average energy per atom must be identical:

$$\overline{E_i} = \frac{1}{N}E = \overline{K_i} + \overline{U}(x_i). \tag{6.5}$$

The difference between average kinetic energy and potential seen in the picture according to the *virial theorem* in classical mechanics is a direct consequence of the anharmonicity of the potential. As is known for harmonic potential, these values are equal to $\overline{K_i} = \overline{U}(x_i)$. The energy $\sim \kappa kT$ accounts for one degree of freedom. This conclusion (*equipartition theorem*) is valid for any kind of ensembles and in space of any dimension κ. The difference between average kinetic energy and potential is responsible for the non-trivial dependence $T(E)$. In turn, it affects an important characteristic of the ensemble, *heat capacity at a constant volume*[3]

$$C_V = \frac{\partial E}{\partial T}. \tag{6.6}$$

According to equation (6.5) this value is proportional to the number of particles. In particular, equations (6.2) and (6.5) for a gas of free particles give

$$C_V = \frac{1}{2}kN, \tag{6.7}$$

This value is one of the *extensive* quantities, whose properties change in proportion with the size (or extent) of the system. In contrast, the temperature is an *intensive* or

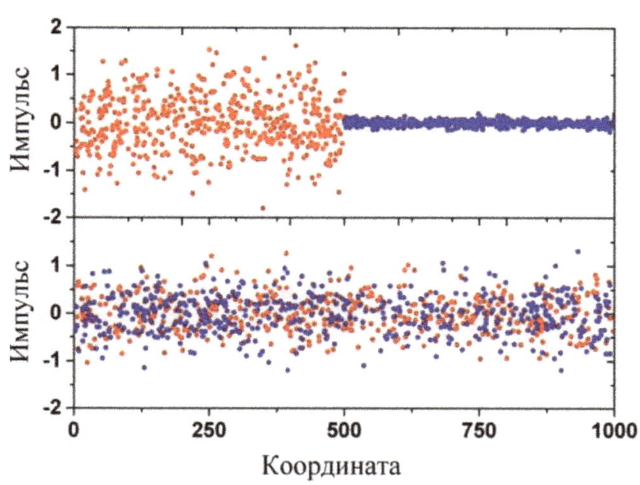

Figure 6.3. The result of mixing of two volumes of gas with different temperatures. Top: before mixing. Below: after mixing, at $t = 10^5$. The red dots are the phase coordinates of the 'hot gas' and the blue dots those of the 'cold gas'.

[3] Recall that all our calculations were made for a constant value of V, so we cannot extend the obtained results to the case of varying volume.

bulk property of a system that does not depend on the system size or the amount of material in the system. To avoid misunderstandings in formulas due to the different system dimensions $\kappa = 1, 2, 3$, it is convenient to introduce a dimensionless heat capacity at constant volume, $c_V = \kappa \frac{C_V}{kN} = \frac{1}{2}, 1, \frac{3}{2}$.

The difference in the heat capacity of an ensemble of interacting particles from its value for an ideal gas shows to what extent the redistribution of total energy between the kinetic and potential energies of the system occurs.

Because of symmetry, it is obvious that any (but rather large) part of a chain is described approximately by the same distribution in the phase space. The only difference from the microcanonical ensemble is due to interaction between parts, breaking slightly the conservation of the partial energy. So the same temperature is maintained in all parts of a chain. The question arises, what will happen if you mix two volumes of an ideal gas at different temperatures?

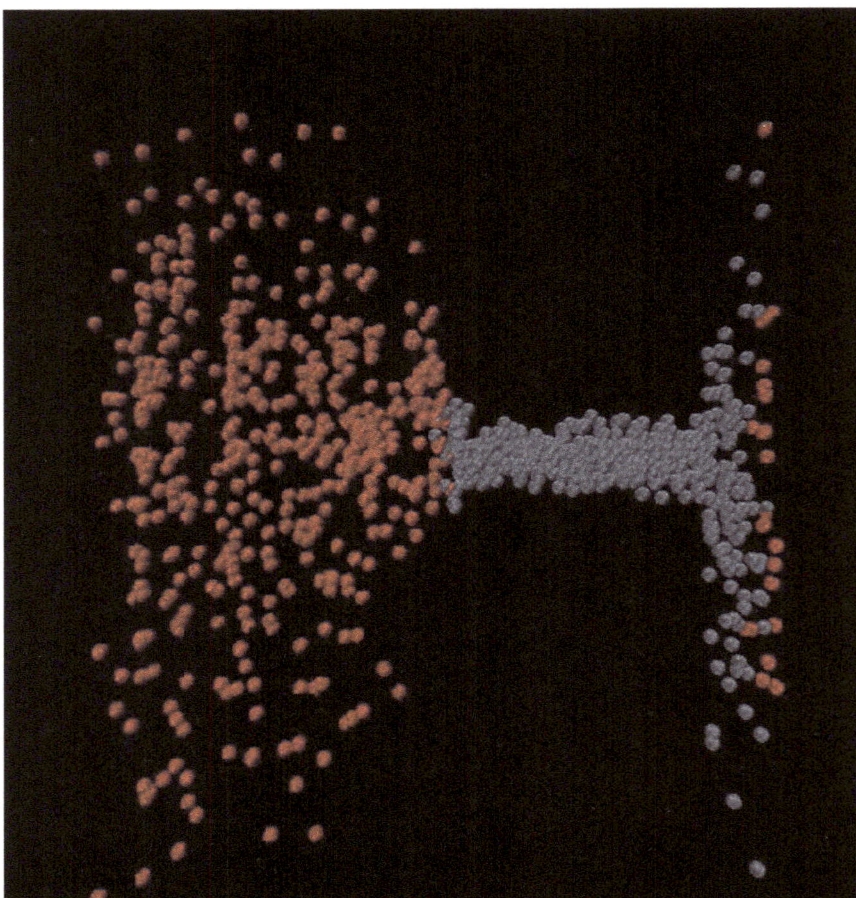

Figure 6.4. Evolution of mixing of two volumes of ideal gas of different temperatures. Animation available at http://iopscience.iop.org/book/978-0-7503-1341-4.

Figures 6.3 and 6.4 show the result of such a calculation for the mixing of two equal volumes of a 1D gas containing $N = 500$ particles at different temperatures T_1 and $T_2 = 0.01\ T_1$ (the details can be found in appendix A.2). After equilibrium in each of them (for this purpose the dynamics was initiated in each volume and traced up to the times $t \sim t_{eq}$), both volumes were adjoined in the point $x = 500$. At time $t \sim 10^4$ complete mixing and the same temperature were established. That is confirmed by the same momentum distribution for both types of particles $\overline{p^2} = 2mE/N$, such that $T_1 > T' > T_2$.

The magnitude of the final temperature T' of the mixture of two volumes of gas consisting of N_1 and N_2 particles, follows from the law of energy conservation

$$NT' = E_1 + E_2 = \frac{N_1}{2}kT_1 + \frac{N_2}{2}kT_2. \tag{6.8}$$

In particular, it follows that at $N_1 \approx N \gg N_2$ the final temperature of the second volume will catch up to the temperature of the first one, called the *thermostat*.

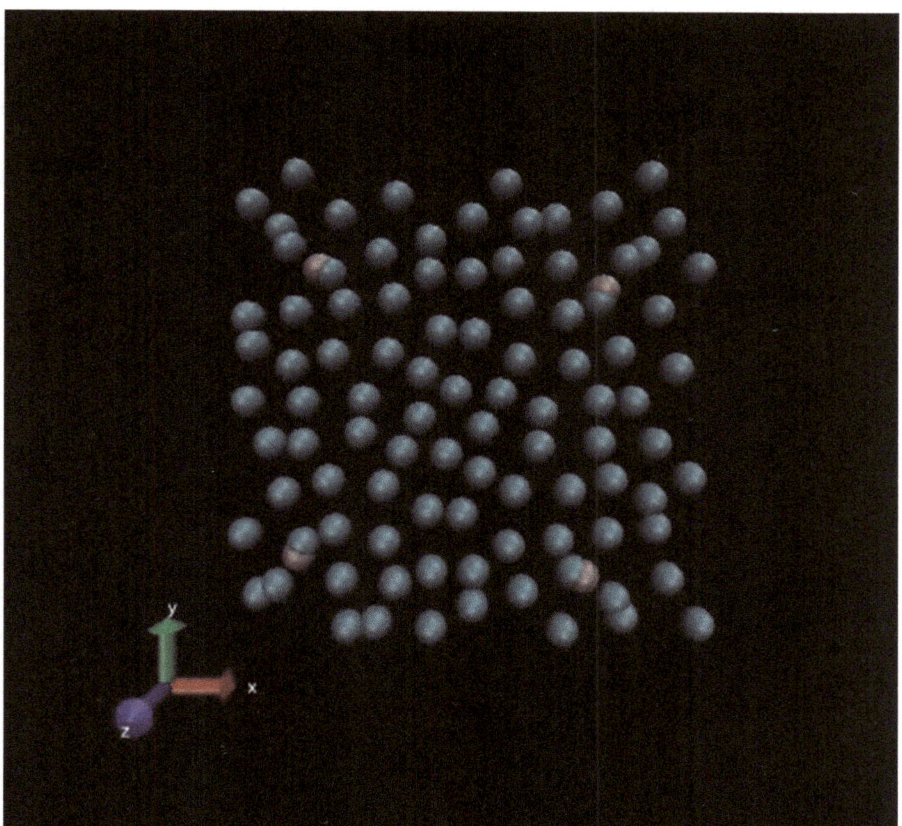

Figure 6.5. Thermalization of a fast particle in a medium. Animation available at http://iopscience.iop.org/book/978-0-7503-1341-4.

A similar phenomenon occurs not only for the two ensembles with different temperatures, but even for a single fast particle incident on a target. After a series of subsequent collisions it is *thermalized* and its kinetic energy becomes $\sim kT$. Such a process can be seen in figure 6.5. After some time, the four particles scattering on other motionless particles become indistinguishable from them. The energy of fast particles is 'exchanged' to the energy of all other particles.

The property of establishment of equilibrium shown in figure 6.3 for two systems interacting with each other indicates the *transitivity of equilibrium*: if two systems are in balance with a third, they are in balance with each other.

A few words should be said about the accuracy of the relationship of equation (6.2) of microcanonical temperature $T_m = (1/mN)\sum_{i=1}^{N} p_i^2$ and absolute temperature. Unlike the latter, as for any microstate the value of T_m fluctuates being distributed via the normal distribution. The situation is well-known in mathematical statistics when the variance is estimated from a sample drawn from a population of random numbers. Thus the credible interval including the exact value of temperature is narrowed with the growth N as $1/\sqrt{N}$ (see problem 6.1) .

At this point it is advisable to introduce an important notion. As we see above, the microcanonical temperature, as do other internal macroscopic parameters of the system below, tends to a constant limit either for $N \to \infty$ or to being averaged over a large interval of time in equilibrium. This limit we call the *thermodynamic limit.* So thermodynamics can be viewed as statistical mechanics in the limit of very large systems.

Problem 6.1. Find the variance of the microcanonical temperature.

Further reading

Dieter H E G 2001 *Microcanonical Thermodynamics: Phase Transitions in 'Small' Systems* (Singapore: World Scientific)

Kittel C 2004 *Elementary Statistical Physics* (New York: Dover)

IOP Publishing

Principles of Statistical Physics and Numerical Modeling

Valeriy A Ryabov

Chapter 7

Pressure, stress and tension

The energy of a closed system occupying a certain volume limited by walls depends on another important macroscopic parameter of a system, its *pressure*. Its difference from zero is revealed by a force acting on the wall although, of course, the pressure is an internal property of a system which has nothing to do with any border forces.

Note that the change of volume is a specific case of a more general phenomenon in condensed matter, deformation. The problem in the introduction of strain and stress in particle dynamics is far from obvious. According to continual mechanics, the forces responsible for deformation are surface or volume valued. Hence, seemingly, they do not reduce to pointwise interatomic forces. To clarify the situation for a microcanonical ensemble, consider a lattice with periodic boundary conditions (PBCs). In fact, the PBCs do not allow the introduction of a variable volume under the influence of a force of any kind. In the absence of thermal vibrations, the force acting on each particle from neighbors (see figure 7.1), is equal to zero, regardless of the state of the system, whether it is squeezed or compressed. Therefore, at first glance, a change of the volume under PBCs may occur only 'by hand'.

Meanwhile, within the model of the chain placed on the circle considered earlier in chapter 4 (see figure 4.1) it is possible to enter the change of volume (the chain length). Take a look at the well-known children's toy the spring (figure 7.2). The squeezed spring is straightened (indicated by arrows), increasing the radius R, and vice versa, the stretched spring tends to return to a smaller radius. The curve of a spring enables us to introduce deformation, compression/extension and reveal the forces responsible for it.

Let us discuss this in more detail. The change in radius $R_0 \rightarrow R$ causes an extension or shortening of the arc coordinate $x = (R/R_0)x_0 = \lambda x_0$ (see figure 7.3). The same *stretch ratio* λ changes the chain length L_0 during the transition from the initial (marked by index zero) to a current configuration: $L = \lambda L_0$. In addition, in the

doi:10.1088/978-0-7503-1341-4ch7

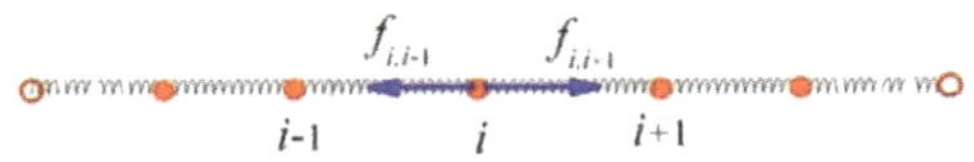

Figure 7.1. Forces in an ideal chain with PBCs. Images are shown at the ends by empty circles.

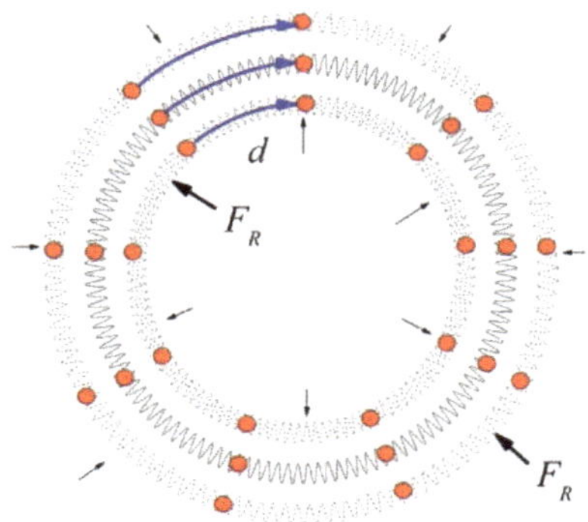

Figure 7.2. On the concept of pressure. The magnitude of compression of the spring is dependent on the radius. The dashed springs are compressed or stretched.

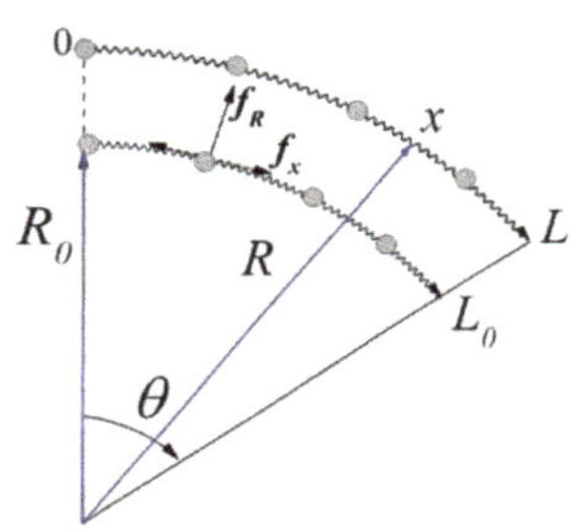

Figure 7.3. Forces in the curved chain. Compression gives rise to the force f_R. The coordinate x changes proportionally to the radius.

polar coordinate system a coordinate of a particle x_i on the circle is measured by the central angle θ_i:

$$x_i = R\,\theta_i. \tag{7.1}$$

As a result, extending the real 1D configuration space to 2D and placing our chain on the circle, we have two dynamical variables. The first coordinate R is responsible for deformation and the second θ_i for the motion along the chain on the circle.

As is clear from figure 7.3, the force f_R causing a change of arc length is directed along the radius. In equilibrium $f_R = 0$ and the radius is constant. Let us calculate this radial force, assuming that the pair potential of interaction between particles depends not on the distance between them, but on the length of the arc connecting them on the circle $x_i - x_{i\pm 1}$. This is necessary for the particle dynamics to be the same as in the real 1D frame given a constant radius. Thus, apart from the interatomic force $f_x = -U'_x$, an additional force f_R acting on each atom of the chain along the radius takes place. To compute this force write the Lagrangian (1.2) in a polar coordinate system and in nearest-neighbor approximation:

$$\begin{aligned}\mathscr{L} &= \sum_{i=1}^{N}(m\dot{x}_i^2 - U(x_i - x_{i-1}))\\ &= \sum_{i=1}^{N}\left(\frac{m}{2}\left(\dot{R}^2 + R^2\dot{\theta}_i^2\right) - U(R(\theta_i - \theta_{i-1}))\right).\end{aligned} \tag{7.2}$$

Then, the generalized force responsible for the radius change and acting on the entire mass Nm is equal to

$$\frac{\partial \mathscr{L}}{\partial R} = \sum_{i=1}^{N}\left[mR\dot{\theta}_i^2 - U'(x_i - x_{i-1})(\theta_i - \theta_{i-1})\right] = Nf_R. \tag{7.3}$$

Equivalently the radial force f_R, shown in figure 7.3, acts on each particle. Now multiplying the numerator and denominator of this expression by R we can return to the variables x_i and $\dot{x}_i = R\,\dot{\theta}_i$ and express the radial force through the stretch ratio λ instead of the radius:

$$f^\lambda = R_0 f_R = \frac{1}{\lambda N}\sum_{i=1}^{N}[m\dot{x}_i^2 - U'(x_i - x_{i-1})(x_i - x_{i-1})]. \tag{7.4}$$

Eventually the force of equation (7.4) does not depend on any artificially introduced geometrical parameters being independent macroscopic characteristics of the system relating to its deformation. Being divided on the initial value of the period $d_0 = L_0/N$, this force gives the tension force (see problem 7.3)

$$\mathsf{T} = \frac{f^\lambda}{d_0} = \frac{1}{L}\sum_{i=1}^{N}\left[m\dot{x}_i^2 + f_{ij}(x_i - x_{i-1})\right], \tag{7.5}$$

where $f_{i,i-1} = -U'(x_i - x_{i-1})$ is the interaction force between the neighboring particles.

We can easily extend the 1D derivation above to the case of a greater number of dimensions thus introducing uniaxial deformation λ_α along each axis X_α. The force of equation (7.4) comes to

$$f_\alpha^\lambda = \frac{\sigma_\alpha}{\lambda_\alpha} = \frac{1}{\lambda_\alpha N}\left(\sum_{i=1}^{N} m\mathrm{v}_{\alpha i}^2 + \sum_{j>i=1}^{N} f_\alpha(r_{ij})\ r_{\alpha ij}\right), \tag{7.6}$$

where $f_\alpha(r_{ij})$ is the α-component of the interaction force between the ith and jth particles.

How the action of this uniaxial tension force generalizes the simple example shown in figure 7.3 is provided as an animation in figure 7.4.

It illustrates the deformation of an ideal 2D lattice along one direction. The deformation along the other perpendicular axis is absent (both directions are marked

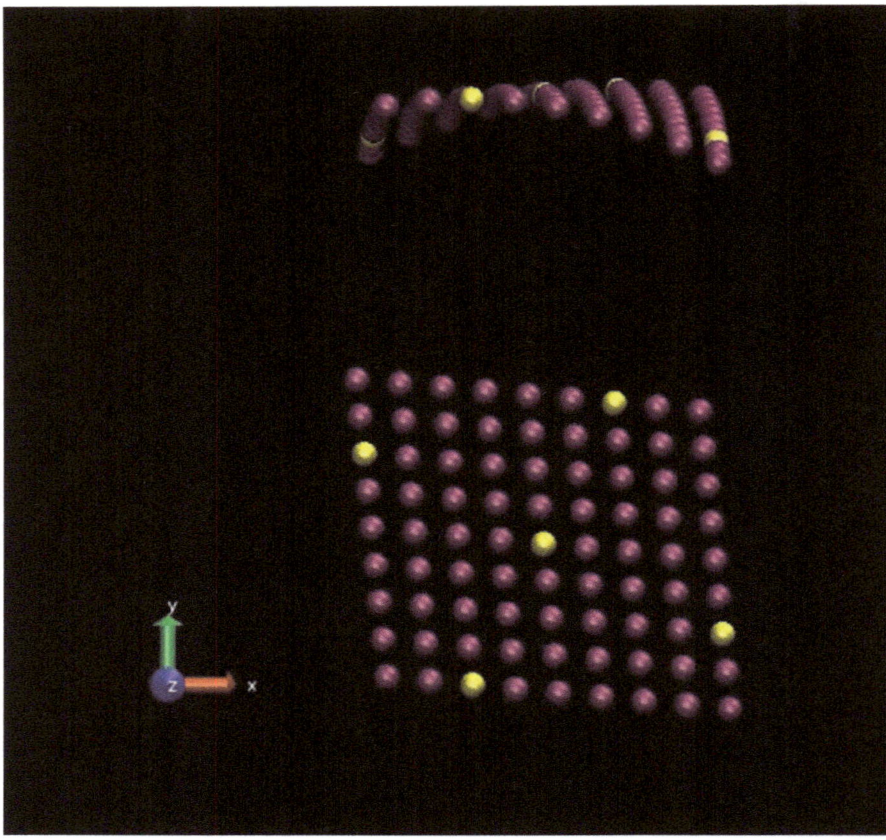

Figure 7.4. Uniaxial strain vibration of a 2D ideal lattice. Animation available at http://iopscience.iop.org/book/978-0-7503-1341-4.

by yellow circles). Therefore, the displacements of coordinates occur perpendicularly to the axis of the cylinder.

Independent stretch along each axis X_α, $\alpha = x, y, z$ causes the change of the volume: $V = \lambda_x\lambda_y\lambda_z V_0$. Assume, for simplicity, that our system occupies a rectangular parallelepiped of the volume $V = L_x L_y L_z$. Then the generalized tension of equation (7.6) corresponds to the tension force analogous to that presented by equation (7.5): $\mathsf{T}_\alpha = Nf_\alpha^\lambda / L_\alpha$. If this force is applied and evenly distributed over the surface element $S_\alpha = V/L_\alpha$ along the same direction, then the value of

$$P_\alpha = \mathsf{T}_\alpha / S_\alpha = \sigma_\alpha / V \tag{7.7}$$

is the so-called *internal uniaxial pressure/stress*. In equilibrium, this pressure equilibrates with traction[1]. Note that in the 1D case the internal pressure of the string coincides with the *tension force* (see problem 7.2).

[1] To avoid misunderstandings we indicate the dimensionality of all these values. Generalized tension = energy, tension force = force, pressure and traction = force/area, traction force = force.

In the isotropic case all stretch ratios and pressures of equation (7.7) are equal: $\lambda_\alpha = \lambda$ and $P_\alpha = P$. Therefore, taking into account $V(\lambda) = \lambda^3 V_0$, we obtain

$$P = \frac{1}{3V}\sum_{\alpha=1}^{3}\sigma_\alpha = \frac{1}{3V}\left(2K + \sum_{j>i=1}^{N} f(r_{ij})r_{ij}\right). \tag{7.8}$$

Since this pressure relates to the reaction of a system to overall compression/extension, it is called *hydrostatic pressure*. Unlike the tension force, this value is associated with the change of energy due to a change of volume. Indeed, one should use a Hamiltonian representation instead of the Lagrangian one. Then, in correspondence with the rules of the classical mechanics (see problem 7.1)

$$\frac{\partial \mathcal{H}}{\partial V} = -\frac{L_\alpha}{V}\frac{\partial \mathcal{L}}{\partial L_\alpha} = -P. \tag{7.9}$$

The value of pressure of equation (7.8) consists of two terms. The first, relating to the doubled kinetic energy K, can be interpreted as the result of momentum transfer from particles to the walls bounding a volume. Let us prove this considering the pressure of free particles, i.e., neglecting the second term in equation (7.8):

$$P = \frac{2K}{3V} = \frac{NkT}{V}. \tag{7.10}$$

Consider a tube of length $v_x\Delta t$ and cross-sectional area A (figure 7.5). Only half of all particles traveling in the tube with velocity $|\mathbf{v}_x|$ reach the wall A in time Δt: $N/2 = \rho v_x \Delta t A/2$, where ρ is the density of particles. Momentum transfer $\Delta p = 2mv_x$ forms the force acting on the wall: $\mathsf{T}_x = \Delta p/\Delta t = (N/2)\,2mv_x/\Delta t = \rho A m v_x^2$. Hence, the pressure $P = \mathsf{T}_x/A = \rho m v_x^2$ along one direction is caused by momentum transfer. Averaging this value over the Maxwell distribution we come to equation (7.10).

The second term, the *virial*, refers to the reaction of continuous media to compression/extension. It is seen from a relation analogous to equation (7.9) and following from equation (7.3) that

$$\frac{\partial \mathcal{H}}{\partial L_\alpha} = -\frac{\partial \mathcal{L}}{\partial L_\alpha} = -\mathsf{T}_\alpha. \tag{7.11}$$

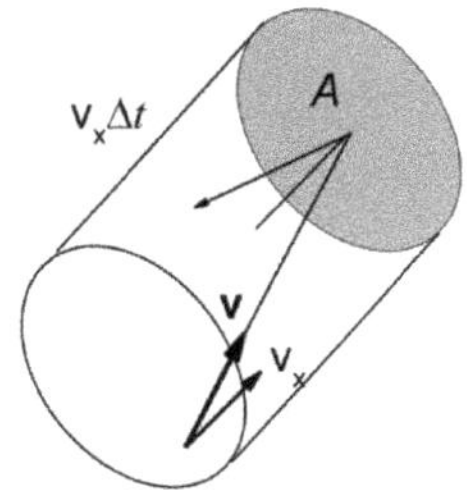

Figure 7.5. Definition of pressure as a momentum flux.

This traction for small deformation in the ideal crystal or in a continuous medium gives rise to a linear dependence on stretching $\mathsf{T} = \text{const} \cdot (L - L_0)$ known as *Hooke's law* (see problem 7.3).

Tension and stress also have an alternative physical sense for condensed matter. These values determine a cohesion force between two halves of a system separated by an imaginary plane (see problem 7.2). This important conclusion provides a link between atomic scale and continuum mechanics calculations, in particular in the problem of an external load (pressure) applied to a body.

In an equilibrium state the tension and the length are preserved. Hence, a certain equilibrium length/volume corresponds to zero tension/pressure for a given temperature. Let us trace the dependence of the pressure on temperature with the help of the model of a 1D chain of particles described in appendix A.3. Recall that temperature in a microcanonical ensemble is defined after achievement of equilibration.

Figure 7.6 shows the dependence of pressure/tension $\overline{\mathsf{T}} = \overline{\mathsf{T}}(d, T)$ on the temperature and the period of the chain calculated according to equation (7.5) (all details of calculation are given in appendix A.6). The derivative of this value over the period gives the elastic constant $\kappa = (\partial \overline{\mathsf{T}} / \partial d)_T$. The inverse value is the 1D analog of *compressibility*

$$\chi = -\frac{1}{V}\left(\frac{\partial V}{\partial P}\right)_T, \tag{7.12}$$

which will be discussed in appendix A.8 in terms of an isobaric ensemble.

The line of intersection of this surface with the base, $\overline{\mathsf{T}}(d, T) = 0$, gives the value of the *thermal expansion coefficient.* It is determined by the slope of the solid line in

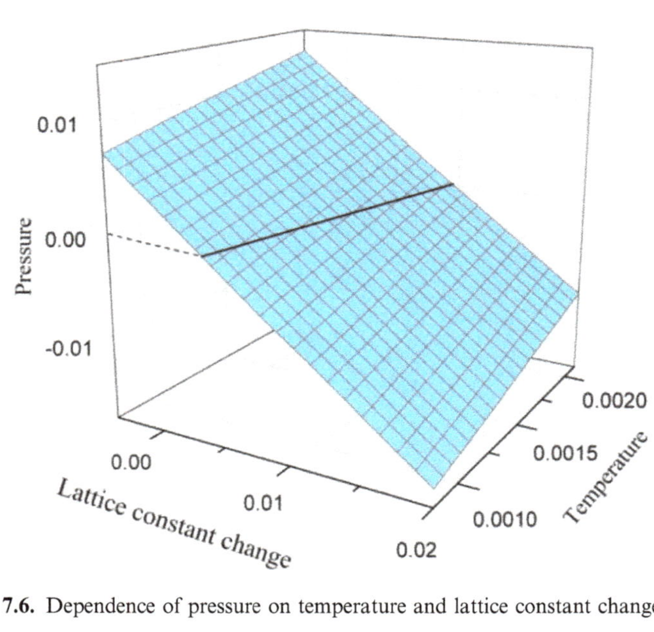

Figure 7.6. Dependence of pressure on temperature and lattice constant change $d - 1$.

figure 7.6. In general, the procedure of pressure vanishing, *optimization*, is often used in molecular dynamics calculations to establish the equilibrium spatial parameters of the system for a given potential (the distances between atoms in molecules, their mutual angular orientation, equilibrium volume and so on). Note that all results in this chapter are the consequence of a microcanonical ensemble, which preserves energy and volume. Below we will consider other types of ensembles for which the volume is a dynamic variable, and the internal pressure is adjusted to a given external pressure.

In conclusion, it should be emphasized that the thermal motion of atoms and all changes in the size or shape of a system are solely a result of interaction between the atoms in 3D configuration space. The approach of extended space and additional dynamical variables suggested here is an approximation based on a strong distinction of the length–time scale related to the thermal motion of atoms and deformation. It allows an easy treatment of many macroscopic values and theorems in statistical mechanics. The choice of the representation for deformation in terms of stretching or volume is a question of convenience, because all results corresponding to them are equivalent.

Problem 7.1. Verify the relation of equation (7.9) using transformation to generalized momentum.

Problem 7.2. Consider an ideal chain in the nearest-neighbor approximation. Prove that for the force of equation (7.5) is equal to the cohesion force between two halves of the chain.

Problem 7.3. Consider an ideal chain in the harmonic approximation. Verify that the tension of equation (7.5) gives rise to Hooke's law.

Further reading

Landau L D and Lifshits E M 1980 *Statistical Physics Course of Theoretical Physics* vol 5 3rd edn (Oxford: Pergamon)

IOP Publishing

Principles of Statistical Physics and Numerical Modeling

Valeriy A Ryabov

Chapter 8

Entropy

In addition to the already introduced macroscopic averages, temperature and pressure, in statistical mechanics another important average is used, entropy. Its origin is directly connected with the ergodic behavior of the system, i.e., with the element of randomness which is formally absent in a purely dynamic description. If the system can be described as random, in terms of the probability distribution, it is an inherent criterion which measures the amount of uncertainty in our knowledge of a system. The quantity of information or, opposite to it in sign, an *entropy* value serves in the probability theory as such a measure. For discrete groups of events with probabilities p_n, it is equal to

$$S = -k\sum\nolimits_n p_n \ln p_n. \tag{8.1}$$

Analogically, for the continuous phase space distribution,

$$S = -k\int f(\mathbf{x}, \mathbf{p}) \ \ln f(\mathbf{x}, \mathbf{p}) \, d\mathbf{x} \, d\mathbf{p} = -k\langle \ln f(\mathbf{x}, \mathbf{p})\rangle. \tag{8.2}$$

The *Boltzmann constant k* is introduced here for convenience, to provide the correct dimension of entropy with respect to other macroscopic averages.

In contrast to the temperature or pressure, the entropy is not a mean value of any mechanical value, although it refers to the whole ensemble. But, if in a simple problem, such as 'whether rain, whether snow, whether is, whether not', where the entropy is equal $S = k\ln 4$, the integral of equation (8.2) is dealing with a very large number of variables and, as a result, with non-trivial dependence of the entropy on energy, temperature and pressure.

To proceed further, it will be convenient to come from the representation in phase space to the energy representation. To perform this transformation, recall that all points in phase space *a priori* are equally probable (see equation (2.5)). Hence, if Γ_E is the number of admissible states of the system with energy E, then $f(E) = 1/\Gamma_E$ and according to equation (8.1) we obtain

doi:10.1088/978-0-7503-1341-4ch8

$$S = -k\,\Gamma_E \frac{1}{\Gamma_E} \ln \frac{1}{\Gamma_E} = k \ln \Gamma_E. \tag{8.3}$$

To express the energy representation with the same in the phase space, introduce the normalized to unity microcanonical distribution $f(\mathbf{r}, \mathbf{p})$ of equation (3.3)

$$f(\mathbf{x}, \mathbf{p}) = \frac{\delta(E - \mathscr{H}(\mathbf{x}, \mathbf{p}))}{\Omega(E)}, \tag{8.4}$$

where

$$\Omega(E) = \int \delta(E - \mathscr{H}(\mathbf{x}, \mathbf{p}))\, d\mathbf{x}\, d\mathbf{p}. \tag{8.5}$$

The integral of this value over the energy from zero to E

$$\Gamma(E) = \int_0^E \Omega(E')dE', \tag{8.6}$$

is, obviously, the *volume of phase space*, for which $H(\mathbf{x}, \mathbf{p}) \leqslant E$.

The derivative of this value, the *density of states*

$$\Omega(E) = \frac{\partial \Gamma}{\partial E}, \tag{8.7}$$

means that $\Omega(E)\, dE$ is the number of states in a thin layer between two surfaces $\mathscr{H}(\mathbf{x}, \mathbf{p}) = E$ and $\mathscr{H}(\mathbf{x}, \mathbf{p}) = E + dE$ (recall in this regard that the volume of phase space in the microcanonical ensemble is equal to zero) .

Integration of equation (8.6) on energy gives

$$\Gamma(E) = \int \vartheta(E - \mathscr{H}(\mathbf{x}, \mathbf{p}))\, d\mathbf{x}\, d\mathbf{p}, \tag{8.8}$$

where $\vartheta(x)$ is the step function, which restricts the integration over the entire space by condition $E - H(\mathbf{x}, \mathbf{p}) \geqslant 0$. Shifting here the kinetic energy to the right-hand side, rewrite this inequality to reveal values of the momenta that are inside the hypersphere of the radius R_p

$$\sum_{i=1}^{N} p_i^2 < R_p^2 = 2m(E - U(\mathbf{x}))\vartheta(E - U(\mathbf{x})). \tag{8.9}$$

Therefore, the integral over the momentum $\mathbf{p}$ in equation (8.8) is equal to the hypersphere volume. It may be set from easy dimensional consideration by analogy with the volume of the usual sphere. At $N = 3$: $\Gamma(E) = \text{const}\, R_p^3$ with an angular constant const $= 4\pi/3$. Omitting larger N, a procedure of calculation of the angular constant in equation (8.8) (it will be excluded from derivation because of a normalization procedure), we straightway have $\Gamma(E) = \text{const}\, R_p^N$ or

$$\Gamma(E) = C \int (E - U(\mathbf{x}))^{\frac{N}{2}} \vartheta(E - U(\mathbf{x}))\, d\mathbf{x}. \tag{8.10}$$

The derivative of this value is

$$\Omega(E) = \frac{CN}{2} \int (E - U(\mathbf{x}))^{\frac{N}{2}-1} \vartheta(E - U(\mathbf{x}))\, d\mathbf{x}. \tag{8.11}$$

The obtained expressions allow us to determine an average of any value $a(\mathbf{x})$ over the density $f(\mathbf{x}, \mathbf{p})$ of equation (8.4) in the form

$$\langle a \rangle = \frac{N}{2\,\Omega(E)} \int a(\mathbf{x})(E - U(\mathbf{x}))^{\frac{N}{2}-1} \vartheta(E - U(\mathbf{x}))\, d\mathbf{x}. \tag{8.12}$$

The angular constant C which has been presented earlier is now excluded.

Use this expression to calculate the entropy. The derivative of the entropy on the energy goes to

$$\frac{\partial S}{\partial E} = \frac{\Omega(E)}{\Gamma(E)}. \tag{8.13}$$

In view of equation (8.12) it can be written as

$$\frac{\partial S}{\partial E} = \frac{kN}{2\langle E - U(\mathbf{x})\rangle}. \tag{8.14}$$

The expression in brackets is the kinetic energy of the system, therefore, recalling the definition of temperature, we obtain

$$\frac{\partial S}{\partial E} = \frac{kN}{2\langle K \rangle} = \frac{1}{T(E)}. \tag{8.15}$$

Integration of this ratio gives entropy

$$S(E) = \int_0^E \frac{dE'}{T(E')}. \tag{8.16}$$

This relation is used for the numerical calculation of entropy, since an inverse dependence of energy $E(T)$ (the *equation-of-state*) is the most frequently requested in calculations.

Returning to the phase space volume of equation (8.10) for a gas of free particles we have

$$\Gamma(E) = C\; V\; E^{\frac{N}{2}}. \tag{8.17}$$

Let us recall here that transition from a 1D case to a 3D case may be carried out by the replacement of $L = Nd$ by volume V and $N \to 3N$ in power.

But before writing from here the final expression for entropy, it is necessary to make one remark. The ratio of equation (8.3) indicates an important property of entropy, its *additivity*. The previously mentioned almost independence of the phase spaces relating to different subsystems, for example, to the halves of a chain whose dynamics was studied in chapter 4, means that $\Gamma(E)$ of equation (8.8) can be represented as the product

$$\Gamma(E) = \Gamma_1(\frac{E}{2})\Gamma_2(\frac{E}{2}), \tag{8.18}$$

with the factors Γ_1 and Γ_2 relating to each half. Taking into account equation (8.3), it follows that the entropy of the whole ensemble is composed of the entropies of its parts

$$S = S_1 + S_2. \tag{8.19}$$

However, substitution of the expression of equation (8.17) into equation (8.18) does not confirm the property of additivity of entropy because the volume V^N is under logarithm in equation (8.3), so that the splitting of the volume $V = V_1 + V_2$ does not lead to the desirable result, in which $S_1 = S_1(V_1)$ and $S_2 = S_2(V_2)$. This circumstance known as the *Gibbs's paradox* can be resolved by means of removal of an uncertainty of a normalization constant. As was already mentioned in the derivation of equation (8.10) and equation (8.17), the volume of the phase space contains some normalizing constant C, which is nowhere specified. Since the normalized distribution is used for calculation of all macroscopic averages, such as, for example, of equation (8.4), the constant C there was reduced. But in the expression for the entropy of equation (8.2), the distribution function is met under the sign of the logarithm. Therefore, this reduction of a constant C is valid only in partial derivatives of the entropy.

The way out of this paradox lies in taking into account the *identity* of the particles, which we did not consider anywhere, but whose effect can be included in constant C. Namely, states in the phase space that differ by only the permutation of particles are to be considered as identical. It must be emphasized here that it is not a matter of the identity of the particles being important in quantum mechanics. Again recall that we are not dealing with a particle system evolving in time, but with the ensemble for which the probability of a microstate resulting from such permutations is the same.

So, ignorance of the real number of states underestimates the volume of phase space $N!$ times. Therefore, it should be necessary to enter this factor into a normalization constant: $C \sim N!$ Taking into account Stirling's formula $N! = N \ln N - N$, from equation (8.17) we obtain

$$S(E) = kN \ln\left[C' \frac{V}{N}\sqrt{\frac{E}{N}}\right] + \frac{1}{2}N, \tag{8.20}$$

where the constant C' does not depend on the volume or on the energy, or on the number of particles. Indeed, this expression is an additive one, since the argument of the logarithm includes the volume and the energy per particle.

The derivative of the entropy on energy E coincides with the result received earlier

$$\frac{\partial S}{\partial E} = \frac{kN}{2E} = \frac{1}{T}. \tag{8.21}$$

It should be noted that this ratio is normally used in textbooks as a definition of the temperature. It becomes unique in systems (for example, spin systems) where the notion of kinetic energy with respect to momenta is absent. Thus, the temperature there is to be determined only from the dependence of entropy on energy (see chapter 22).

Let us now define the derivative of entropy on volume at constant energy using the equations of (8.8) and (7.9). Changing the order of differentiation in the integrand we obtain

$$\left.\frac{\partial \Gamma}{\partial V}\right|_E = -\left.\frac{\partial}{\partial E}\right|_V \int \vartheta(E - \mathscr{H}(\boldsymbol{x}, \boldsymbol{p}))\frac{\partial \mathscr{H}}{\partial V} d\boldsymbol{x}d\boldsymbol{p} = \frac{P}{T}\Omega. \tag{8.22}$$

From this follows the relation

$$\frac{\partial S}{\partial V} = \frac{P}{T}, \tag{8.23}$$

establishing the dependence of the entropy on pressure. It should be stressed that in contrast to the mechanistic notion of the pressure introduced earlier in chapter 7, here it is a property of phase space revealing a response of pressure to the decrease/increase of integration limits of spatial coordinates. Hence it has nothing to do with any sophisticated geometry in extended space.

Furthermore, it is more convenient to not use the absolute value of entropy of equation (8.20), but the change in entropy when going from one state to another state

$$\Delta S = \int_{S_0}^{S} dS = \int_{T_0}^{T} \frac{\partial S}{\partial T} dT + \int_{V_0}^{V} \frac{\partial S}{\partial V} dV. \tag{8.24}$$

Using equations (8.21), (8.23) and (7.10) for the gas of any dimensionality we obtain

$$\Delta S = c_V Nk \ln\left(\frac{T}{T_0}\right) + Nk \ln\left(\frac{V}{V_0}\right). \tag{8.25}$$

Returning to the results of chapters 1–5, where we observed the transition from purely dynamic to chaotic behavior of the system, it is advisable to raise the question of entropy evolution in this process. The answer relates to another important property of entropy, following from its dependence on time. It is clear that a transition of this kind must be accompanied by a monotonic increase in the volume of the phase space where ordered motion gives way to chaos. Consequently, it leads to a monotonic growth of entropy with time. We underline that the final point of *entropy increase* is achieved when all states in the system become equiprobable (see problem 8.1).

Problem 8.1 A system occupies one of $i = 1 \div N$ states with probability p_i. Show, using the method of the Lagrange multiplier, that the maximum of the entropy corresponds to the probability distribution $p_i = 1/N$.

Problem 8.2. Find the entropy change when two volumes of ideal gases containing N_1 and N_2 molecules, at the same temperature and pressure, are mixing (entropy of mixing).

Further reading

Attard P 2002 *Thermodynamics and Statistical Mechanics: Equilibrium by Entropy Maximisation* (New York: Elsevier)

Penrouse O 1976 *Foundation of Statistical Mechanics* (Oxford: Pergamon)

IOP Publishing

Principles of Statistical Physics and Numerical Modeling

Valeriy A Ryabov

Chapter 9

Gibbs distribution

The introduction of the microcanonical ensemble was based on the law of conservation of energy of a closed system inherent to a certain number of particles. Knowledge of the trajectories of motion allowed us to solve some problems, however, one central problem remained: what is the reason for the Gauss-like behavior of the coordinate and momentum distributions? It turns out that it could be solved by using another ensemble, which uses not energy conservation but the almost complete independence of the behavior of parts of the whole system. In fact, we already encountered this ensemble in chapter 6 when we discussed the thermostat, a large system whose energy and number of orders of freedom essentially overestimate those of another system in contact. The latter means that the microcanonical temperature of a thermostat with $N \to \infty$ is a constant, while the microcanonical temperature and the total energy of a small ensemble fluctuate (see problem 6.1).

Ignoring energy conservation, the solution of the Liouville equation can be written in terms of an arbitrary function of equation (3.2)

$$f(\mathbf{x}, \mathbf{p}) = w(\mathcal{H}(\mathbf{x}, \mathbf{p})). \tag{9.1}$$

Although this result was obtained for a closed system, we have already seen that a fairly large part of the ensemble should be characterized by the same universal distribution which is independent of the initial one. In particular, two equal halves of a chain each have a total energy $\simeq E_0/2$ exchanged between them by energy of order E_0/N. This means that these halves are almost independent, as well as the distribution of their phase coordinates (as this leads to entropy additivity). Statistical independence of the distributions of both parts (which we label with indices 1 and 2) means that

$$f(\mathbf{x}, \mathbf{p}) = w(\mathcal{H}(\mathbf{x}^1, \mathbf{p}^1))w(\mathcal{H}(\mathbf{x}^2, \mathbf{p}^2)), \tag{9.2}$$

wherein

doi:10.1088/978-0-7503-1341-4ch9

$$H(\mathbf{x}, \mathbf{p}) \simeq H(\mathbf{x}^1, \mathbf{p}^1) + H(\mathbf{x}^2, \mathbf{p}^2). \tag{9.3}$$

It follows that the function w is exponential with some universal constant β: $w \sim \exp(-\beta H(\mathbf{x}, \mathbf{p}))$.

But the form of the momentum distribution which is the integral of the function $f(\mathbf{x}, \mathbf{p})$ on coordinates was already established by equation (6.3):

$$\int f(\mathbf{x}, \mathbf{p})\, d\mathbf{x} \sim \exp\left(-\beta \sum_{i=1}^{N} \frac{p_i^2}{2m}\right) = \exp(-\beta K). \tag{9.4}$$

Therefore, the function w in equation (9.2) has the form

$$f(\mathbf{x}, \mathbf{p}) \sim \exp(-\beta \mathcal{H}(\mathbf{x}, \mathbf{p})) \tag{9.5}$$

with $\beta = 1/kT$.

However, equation (9.4) was obtained by the numerical simulation of ideal gas and, consequently, needs verification from a general principle. Since atoms in ideal gas do not interact with each other they can be considered as statistically independent. It means that the joint probability of a system of N atoms energy $E = \sum_{i=1}^{N} \varepsilon_i$ is equal to the product of the separate probabilities $p(\varepsilon_i)$:

$$P(E) = \prod_{i=1}^{N} p(\varepsilon_i). \tag{9.6}$$

Take the differential of the logarithm of both parts of this equation

$$\frac{P'(E)}{P(E)} dE = \sum_{i=1}^{N} \frac{p'(\varepsilon_i)}{p(\varepsilon_i)} d\varepsilon_i. \tag{9.7}$$

Because of the arbitrariness of $d\varepsilon_i$, all coefficients before the differentials must be equal. Hence

$$\frac{p'(\varepsilon_i)}{p(\varepsilon_i)} = -\beta. \tag{9.8}$$

Since the average kinetic energy $\langle E \rangle = NkT$, then $\beta = 1/kT$ and the integration of this equation gives (9.4).

Divided by a normalizing factor, the *partition function*

$$Q = \int \exp(-\beta \mathcal{H}(\mathbf{x}, \mathbf{p}))\, d\mathbf{x}\, d\mathbf{p}, \tag{9.9}$$

the distribution

$$f(\mathbf{x}, \mathbf{p}) = \frac{1}{Q} \exp(-\beta \mathcal{H}(\mathbf{x}, \mathbf{p})) \tag{9.10}$$

is known as the canonical Gibbs distribution.

Sometimes the energetic representation of this distribution could be more convenient. A direct integration of Q of equation (9.9) on phase space with an equidistant step on coordinate and momentum gives the set of energies $E_n = \mathcal{H}(\mathbf{x}_i, \mathbf{p}_j)$. Then with accuracy up to constants

$$Q = \sum_n \exp(-\beta E_n) \tag{9.11}$$

and correspondingly the Gibbs distribution in *energetical representation* becomes

$$f(E) = \frac{1}{Q} \exp(-\beta E). \tag{9.12}$$

Integral representation of the partial function of equation (9.11)

$$Q = \int dEg(E)\exp(-\beta E) \tag{9.13}$$

is expressed through the density of states $g(E)$ normalized on the unit. Warning: equations (9.9) and (9.13) have different dimensions.

What is the difference between the canonical and microcanonical ensembles? In the microcanonical ensemble, the microcanonical temperature fluctuates so the distribution of equation (6.2) itself is defined by a random variable (kinetic energy per atom). As was already noted, this situation is well-known in mathematical statistics. In contrast, the temperature in the canonical ensemble is an external parameter causing fluctuations of the kinetic and total energy. With an increase in N, fluctuations become negligible and the difference between the two types of ensembles disappears and their averages tend to the same thermodynamical values. However, the above is not all so evident. The microcanonical distribution has a sharp maximum while the canonical distribution of equation (9.12), seemingly, gradually decreases with E. The reason for this difference lies in the density of states near a mean energy of the system E_0. The average of any value of equation (9.12) (or, equivalently, the summation on E_n) apart from the distribution includes the factor density of states $\sim e^{-\beta E}\Omega(E)$.

The first multiplier in this product, the exponent, is a rapidly decreasing function of E. The second multiplier $\Omega(E)$ is a rapidly increasing function with the increase of E. For example, if you take equation (8.11) for an ideal gas where $\Omega(E\char`~)E^{\frac{N}{2}-1}$, then the product of these two functions gives a distribution with a sharp maximum, as is shown in figure 9.1. Thus the canonical distribution tends to the microcanonical distribution $f(E) = \delta(E - E_0)$ in limit $N \to \infty$ as it must be.

Now turn to the Gauss-like coordinate distribution obtained in the simulations of chapter 4. Since integration on the momenta in equation (9.10) can be carried out in full, the configuration part of the statistical integral (the *Boltzmann distribution*)

$$\widehat{f}(\mathbf{x}) = \frac{1}{\widehat{Q}} \exp(-\beta U(\mathbf{x})) \tag{9.14}$$

can be used for the calculation of any spatial averages. Here $\widehat{Q}(\beta)$ is a normalization constant.

In principle, the potential in this equation may be referred to as a system of only one particle in a thermostat with the given temperature. The animation in figure 9.2 shows the time evolution of such a particle of mass $m = 1$ in a parabolic potential.

The thermostat is assumed as an external system of particles of mass $10^{-3}m$ whose momenta are distributed with a normal distribution with the variance $\delta p = 0.1$. The thermostat particles are scattered on the oscillator in times $t_k = k\Delta t$ ($\Delta t \approx 7.0$). The initial position of the oscillator was on the bottom of potential well changes due to collisions with particles of the thermostat. It is seen how the energy of the oscillator gradually increases due to collisions up to an equilibrium value ~kT. The Gauss form of coordinate distribution follows from equation (9.14) (see also problem 7.3). Note that the oscillator energy is fluctuating in a wide range although its mean value is equal to kT.

Let us prove now the relationships found in calculations of average kinetic energy of equation (6.2) and the property of the lack of correlations between momenta:

$$\frac{1}{m}\langle p_i\, p_j\rangle = \left\langle p_i\,\frac{\partial H}{\partial p_j}\right\rangle = kT\delta_{ij}. \tag{9.15}$$

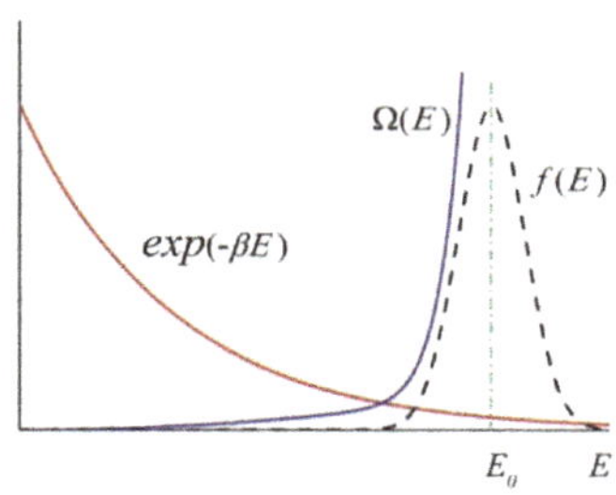

Figure 9.1. Energy representation of the canonical distribution.

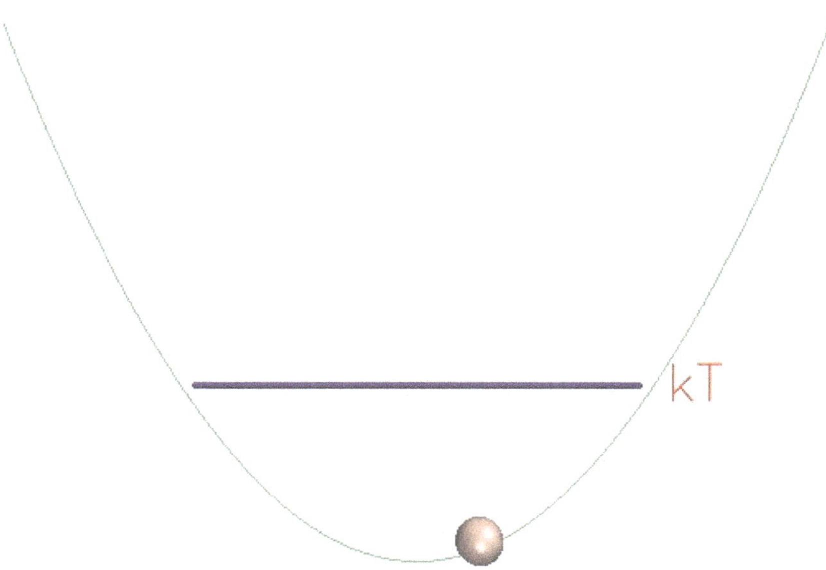

Figure 9.2. Brown oscillator in a thermostat. Animation available at http://iopscience.iop.org/book/978-0-7503-1341-4.

It follows from equation (9.10) that

$$\left\langle p_i \frac{\partial H}{\partial p_j} \right\rangle = \frac{1}{Q} \int p_i \frac{\partial H}{\partial p_j} e^{-\beta H(\mathbf{x},\, \mathbf{p})}\, d\mathbf{x}\, d\mathbf{p}. \tag{9.17}$$

While separating the total derivative in the integrand

$$p_i \frac{\partial e^{-\beta H(\mathbf{x},\, \mathbf{p})}}{\partial p_j} = \frac{\partial}{\partial p_j}\left(p_i\, e^{-\beta H(\mathbf{x},\, \mathbf{p})}\right) - e^{-\beta H(\mathbf{x},\, \mathbf{p})}\delta_{ij}, \tag{9.18}$$

write equation (9.17) as

$$\begin{aligned} \left\langle p_i \frac{\partial H}{\partial p_j} \right\rangle &= -\frac{1}{\beta Q} \int d\mathbf{x}\, d\mathbf{p} \frac{\partial}{\partial p_j} p_i e^{-\beta H(\mathbf{x},\, \mathbf{p})} + \frac{1}{\beta Q}\delta_{ij} \\ &\quad \int d\mathbf{x}\, d\mathbf{p}\, e^{-\beta H(\mathbf{x},\, \mathbf{p})} \\ &= \int d\mathbf{p}\, p_i e^{-\beta H(\mathbf{x},\, \mathbf{p})}\Big|_{p_j=-\infty}^{\infty} + kT\delta_{ij} = kT\delta_{ij}. \end{aligned} \tag{9.19}$$

The found relation, one of the *virial theorems*, could be similarly proved with the help of the microcanonical distribution. In addition, it is easy to repeat the derivation above for coordinates as well:

$$\left\langle x_i \frac{\partial H}{\partial x_j} \right\rangle = kT\delta_{ij}. \tag{9.20}$$

However, unlike equation (9.15) this ratio does not relate to the correlation between the coordinates (see problem 9.2).

The important point of this proof is that the virial theorem for the momentum stands whether there is an interaction between particles or not.

Problem 9.1. Show that a state with maximum entropy for a 1D gas of free particles (see chapter 4) in equilibrium corresponds to the Maxwell distribution.

Problem 9.2. Find the correlation coefficient of atomic displacements in a chain with the nearest-neighbor interaction between them.

Further reading

Cowan B 2005 *Topics in Statistical Mechanics* (Singapore: World Scientific)

Levich B G 1971 Electromagnetic processes in matter *Theoretical Physics: An Advanced Text: Statistical Physics* vol 2 (New York: Elsevier)

Chapter 10

Isothermal ensemble

If the temperature is an external parameter, as was discussed in a previous chapter, how does it remain prescribed constant in the microcanonical ensemble? Recall that in the dynamics the microcanonical ensemble temperature is determined as a result of established equilibrium and therefore cannot be predicted precisely in advance.

The simulation procedure for an *isothermal ensemble* with the maintenance of temperature in the course of evolution requires a change in the velocity of particles in such a way that the average kinetic energy per one degree of freedom on average must be equal to $kT/2$. From classical mechanics, it is known that this role can be performed by friction, guided by a term on the right-hand side of the equation of motion of (1.1), being proportional to the velocity $\tilde{\gamma}\dot{x}_i$. The sign of value γ defines braking ($\gamma < 0$) or acceleration processes ($\gamma > 0$) of the velocity of a particle, causing the kinetic energy to be decreased or increased. Apparently, an additional dynamical variable should be responsible for a change γ in time evolution. However, the procedure of its inclusion in the motion equations, as a *thermostat,* does not seem trivial. Any candidate for the modification of the equations of motion of MD has to: (a) be reversible in time; (b) give an unbiased estimate (even of order of ~$1/N$) of the mean values of the conserving dynamic values, the integrals of the movement, such as energy, and total impulse of the system, and (c) satisfy the virial theorems.

The idea of the most commonly used method (the *Nosé–Hoover method*) meeting all listed requirements consists in the inclusion of the additional variable ξ in dynamics to scale the time t: $d\tilde{t} = \xi dt$ and thus to change the particle velocity at invariable coordinates:

$$\dot{\tilde{x}} = \xi^{-1}\dot{x}. \tag{10.1}$$

Marking all variables relating to the scaled time by 'cap', we also introduce scaled variable $\tilde{\xi}$ by means of the obvious relation $\dot{\tilde{\xi}} = \tilde{\xi}^{-1}\dot{\xi}$. If Q is the 'mass' of the thermostat for the new degree of freedom $\tilde{\xi}$, the Lagrangian of equation (1.2) can be rewritten as

doi:10.1088/978-0-7503-1341-4ch10

$$\mathscr{L} = \frac{m}{2}\sum_i \xi^2 \dot{\tilde{x}}_i^2 - U + \frac{1}{2}NQ\dot{\tilde{\xi}}^2 - NkT\ \ln \tilde{\xi}. \tag{10.2}$$

The first term here is the kinetic energy of the system after time scaling, the third term is the kinetic energy of a new variable $\tilde{\xi}$. The last term takes the role of the 'potential' for a variable ξ, introduced so that the force $\partial L/\partial \tilde{\xi}$ acting on the thermostat becomes proportional to the difference between the kinetic energy of the particles and the thermodynamical limit $NkT/2$. It allows the system to maintain the desired temperature T.

Now return to the previous velocities and time t. Inverse transformation of variables $\tilde{\xi}$, $\tilde{x} \to \xi$, x gives

$$\begin{aligned} \dot{\xi} &= \xi\dot{\tilde{\xi}}, \quad \ddot{\xi} = \tilde{\xi}^2\ddot{\tilde{\xi}} + \xi\dot{\tilde{\xi}}^2, \\ \dot{x} &= \xi\dot{\tilde{x}}, \quad \ddot{x} = \xi^2\ddot{\tilde{x}} + \xi\dot{\tilde{x}}^2. \end{aligned} \tag{10.3}$$

Substituting these relations into Euler's equations of (1.3), and introducing a new variable $\gamma = \dot{\xi}/\xi$ into them instead of ξ, we eventually come to the motion equations

$$\begin{aligned} \ddot{x}_i &= \frac{1}{m}F_i - \gamma\dot{x}_i, \\ Q\ddot{\gamma} &= \frac{2}{N}K - kT. \end{aligned} \tag{10.4}$$

Here $\dot{\gamma}$ denotes the velocity of the thermostat in which the *mass of the thermostat Q* controls the rate at which the kinetic energy of the system adjusts to a given temperature. In specific calculations, it should not be too small, otherwise the phase space of the system ceases to satisfy the canonical distribution, but also not too large—in this case, kinetic energy will slowly adjust to the temperature. In practice, the process of thermostatting includes the use of a chain of several thermostats to control the temperature more effectively. In addition, it resolves some problem with phase space (see below).

A direct check ensures that the motion equations of (10.4) preserve the energy of a full system, including a heat bath

$$E = \frac{m}{2}\sum_{i=1}^{N}\dot{x}_i^2 + \frac{N\dot{\gamma}^2}{2Q} + U + NkT\gamma \tag{10.5}$$

while the energy of a physical system $K + U$ fluctuates.

The structure of the equation of motion of (10.4) suggests the attachment of the thermostat to any independent order of freedom. Hence it can be extended to an ensemble of any dimension without change. Figure 10.1 illustrates the solution of the equations of motion of (10.4) for a 2D lattice with the different thermostat masses Q. Methods of a solution and details of the simulation model are presented in appendix A.7. We will also become acquainted with other examples of the use of isothermal ensembles in later chapters.

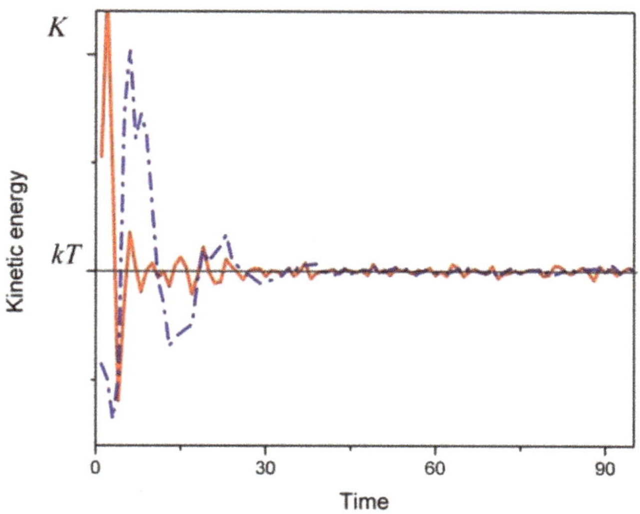

Figure 10.1. Evolution of the kinetic energy in an isothermal ensemble. The red curve corresponds to thermostat mass Q ten times lower than for the blue curve.

In conclusion, we note that the considered motion equations belong to so-called non-Hamilton dynamics, for which many theorems concerning Hamilton systems are not applicable. The proof of the consistency of the approach developed above can be found in the references below.

Problem 10.1. Prove that the stationary state of a system with a Nosé–Hoover thermostat obeys the canonical distribution.

Further reading

Hoover W G 1985 *Phys. Rev.* A **31** 1695

Nose S 1984 *J. Chem. Phys.* **81** 511

IOP Publishing

Principles of Statistical Physics and Numerical Modeling

Valeriy A Ryabov

Chapter 11

Free energy

One of the major problems in the classical mechanics of systems of many particles is the determination of the minimum of a potential. The position of the minimum determines the equilibrium of a mechanical system with respect to macro parameters referring to a system. Therefore, it seems quite natural to ask the question: what replaces the condition of a minimum of mechanical energy, when we consider an ensemble of states with constant temperature and pressure? An analog of potential energy in this case is the so-called *free energy*[1]

$$F = -\frac{1}{\beta}\ln Q = -\frac{1}{\beta}\ln\left(\int \exp(-\beta\mathscr{H}(\mathbf{x}, \mathbf{p}))d\mathbf{x}\,d\mathbf{p}\right). \tag{11.1}$$

Why the logarithm? It turns out that we have to use exactly this value for the calculation of average values relating to the canonical ensemble and it is already known to us from the microcanonical ensemble. So the average of energy

$$E = \langle\mathscr{H}\rangle = \frac{1}{Q}\int \mathscr{H}\, e^{-\beta\mathscr{H}}d\mathbf{x}\,d\mathbf{p} = -\frac{\partial}{\partial\beta}\ln Q. \tag{11.2}$$

Further, for the pressure of equation (7.9), we have

$$\langle P\rangle = -\left\langle\frac{\partial\mathscr{H}}{\partial V}\right\rangle = kT\,\frac{\partial\ln Q}{\partial V} = -\frac{\partial F}{\partial V}. \tag{11.3}$$

You have to keep in mind (see chapter 8) that now the pressure is a property of phase space but not a consequence of the dynamics in extended space. Therefore quite an analogous relation for uniaxial tension coexists with equation (11.3):

$$\left\langle f^l_\alpha\right\rangle = -\frac{1}{N}\frac{\partial F}{\partial\lambda_\alpha}. \tag{11.4}$$

[1] To avoid confusion in the future, we will omit the constant factor ~$1/N!$ in front of the integral (see the discussion in chapter 8).

doi:10.1088/978-0-7503-1341-4ch11

From (11.1) it follows that that $Q = e^{-\beta F}$. Hence the phase density

$$f(\boldsymbol{x}, \boldsymbol{p}) = e^{\beta(F-\mathscr{H})}. \tag{11.5}$$

Or in the energetical representation of equation (9.12)

$$f(E) = e^{\beta(F-E)}. \tag{11.6}$$

Substituting it in equation (8.2), we obtain

$$S = -\frac{1}{T}\int (F - E)\, e^{\beta(F-E)} dE. \tag{11.7}$$

The result of averaging comes to

$$F = E - TS. \tag{11.8}$$

Differentiation of free energy on temperature taking into account $F = E - TS$ and also $\partial\beta/\partial T = -k\beta^2$ gives

$$S = -k\frac{\partial F}{\partial T}. \tag{11.9}$$

Differentiating this expression on E, we receive the same result, as for a micro-canonical ensemble

$$\frac{\partial S}{\partial E} = \frac{1}{T}. \tag{11.10}$$

And, finally, the heat capacity

$$C_V = \frac{\partial E}{\partial T} = \frac{\partial E}{\partial \beta}\frac{\partial \beta}{\partial T} = k\beta^2 \frac{\partial E}{\partial \beta^2} \ln Q = \beta^2 \frac{\partial F}{\partial \beta^2}. \tag{11.11}$$

Taking into account equations (11.3) and (11.10), the total differential of energy takes the form

$$dE = T\, dS - P\, dV. \tag{11.12}$$

Similarly,

$$dF = d(E - T\, S) = -S\, dT - P\, dV. \tag{11.13}$$

This ratio answers the question at the beginning of this section. The derivative of free energy on time,

$$\frac{dF}{dt} \leqslant -S\frac{dT}{dt} - P\frac{dV}{dt} = 0, \tag{11.14}$$

is less than zero for a closed system under constant temperature and volume. It is known as the *Helmholtz free energy minimum principle.* In other words, a system tends to equilibrium to be defined by a minimum of free energy.

In the expression of equation (11.1) it is possible to perform integration on impulses of particles just as was done in an expression for the volume of phase space equation (8.10). Selecting the factor $\exp(-\beta K(\boldsymbol{p}))$ in the exponent under the sign of a logarithm and carrying out integration on $\mathbf{p}$, we obtain

$$F = F_{id} - kT \ln \left\{ \frac{1}{V^N} \int e^{-\beta U(\boldsymbol{x})} d\boldsymbol{x} \right\}. \tag{11.15}$$

Here the free energy for ideal gas F_{id} consists of the entropy of equation (8.20) and kinetic energy $E = K = \frac{3}{2} NkT$.

Accurate calculation of free energy for a condensed substance is nearly impossible due to insufficient sampling in a finite length and time scale simulation (see problem 11.2). However, nothing prevents us from an integration of the differential relations obtained above. It is advisable not only for the sake of simplification of calculations. Free energy is determined up to a constant term because the integral in equation (11.1) has a dimension of action. Hence only the difference in free energy makes sense. Unlike mechanical energy, this quantity is neither intensive nor extensive. If we combine two identical systems, we will not have twice the free energy (see the Gibbs mixing paradox in chapter 8).

As an example of a method of computation of free energy, carry out the integration of equation (11.3) on volume

$$F(V_2) - F(V_1) = -\int_{V_1}^{V_2} P(V) dV. \tag{11.16}$$

The pressure of the system of equation (7.8) is determined from constant temperature simulations (see the previous section) at a series of volumes between V_1 and V_2, and the integral here can be evaluated numerically.

Figure 11.1 shows the result of such a calculation for a system of $N = 4000$ atoms interacting via the Lennard-Jones potential in an *FCC* lattice (described in appendix A.9). Other details of the calculations are in appendix A.8. The profile of potential U of the ideal *FCC* lattice ($P = 0$ and $T = 0$) is seen in the bottom graph. The minimum of the potential corresponds to the magnitude $U_{\min} = U(d_0 = 1.0) = -7.76$ eV. The change of the volume (expansion), is evaluated from the minimum position in units of stretching ratio $(V/V_0)^{1/3} - 1$. The dependence of pressure in MPa (the middle graph of the figure) was used for the calculation of the free energy increment ΔF (top graph) of equation (11.16). In contrast to observed mechanical equilibrium, the gradual decrease of free energy evidences the absence of any equilibrium substance in this range of atomic density. It indicates, confirmed by MD calculation of the time evolution of a structure (see figure 11.2), a process of melting (so the lattice constant refers to an average distance between atoms).

The presented method of integration on thermodynamical parameters is widely used for free energy calculations.

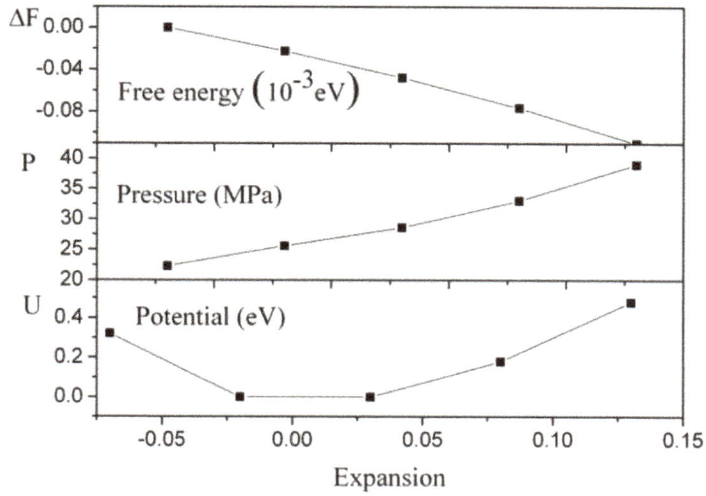

Figure 11.1. Increment of free energy.

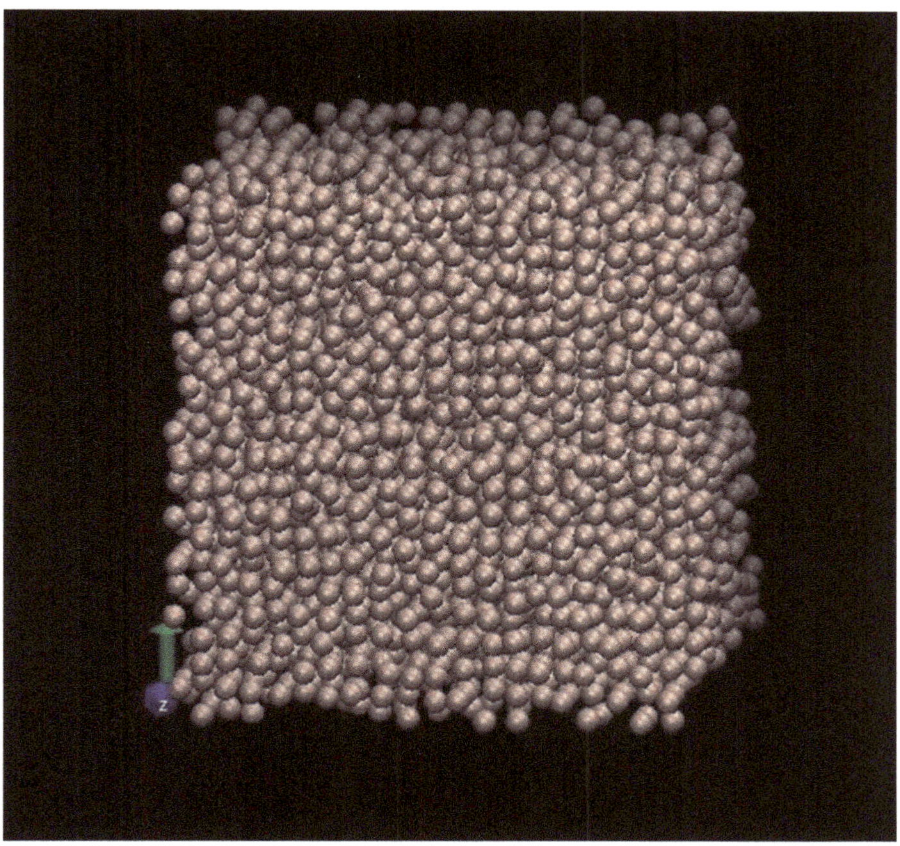

Figure 11.2. Melting process in a 2D system. Animation available at http://iopscience.iop.org/book/978-0-7503-1341-4.

Problem 11.1. Solve the problem 6.1, given that the average value $E = \sum_{i=1}^{N} p_i e_i$.

Problem 11.2. Find the temperature dependence of free energy of the chain of three anharmonic oscillators with PBCs, the period $d = 1$ and the pair potential $U(\Delta) = \frac{1}{2}\Delta^2 + \frac{1}{4}\Delta^4(\Delta = x_i - x_{i-1} - d)$ in the temperature interval $kT = 0.001 \div 0.01$.

Further reading

Chipot C and Pohorille A 2007 *Free Energy Calculations* (Berlin: Springer)

IOP Publishing

Principles of Statistical Physics and Numerical Modeling

Valeriy A Ryabov

Chapter 12

Grand canonical ensemble

As we have seen, a canonical ensemble describes the possible states of part of a system that is in thermal equilibrium with a reservoir. Their interaction may occur not only by an exchange of energy, but also, in principle, an exchange of particles. Although the Hamiltonian in the Gibbs distribution of equation (9.10) relates to a predetermined number of particles, we intentionally omitted the index of this number in all formulas following from a Gibbs distribution, assuming that there could be a different number of particles in a given volume not significantly changing the averages. This is clarified in figure 12.1, where part of a large closed ensemble is shown and whose average values are expected to be the same as those in the reservoir.

First of all, considering an ensemble of N particles filling the volume V, how many particles are in the volume V_1 (marked in black) and what is the free energy of such a subsystem? The average number of particles is then, obviously, equal to $\bar{n} = \dfrac{V_1}{V}N$. To obtain the distribution of p_n, we note that the probability of a particle being present at any point is equal to $p = 1/V$. Thus, the standard combinatory consideration for the probability of falling into the box containing n particles leads to a binomial distribution:

$$p_n = C_{NP}^n p^n (1 - p)^{N-n}. \tag{12.1}$$

In figure 12.2 the distribution of the number of particles in volume $V_1 = 0.2\ V$ is presented. It can be seen that the distribution, even at not very large N, is close to the normal distribution with $\bar{n} = np$. Indeed, at large N the binomial distribution approaches the normal distribution with the variance $\sim 1/\sqrt{N}$. Hence all thermodynamic values are mainly dependent on some average number of particles which is determined self-consistently by the same set of parameters.

Further, since entropy is an additive function, the energy is a function of the entropy and volume per particle $E = Nf(S/N,\ V/N)$. Hence joining of one particle

doi:10.1088/978-0-7503-1341-4ch12

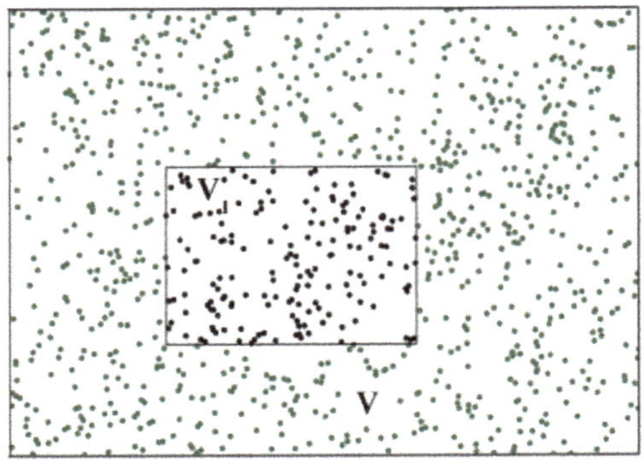

Figure 12.1. The coordinates of atoms. In the volume V_1 they are marked in black).

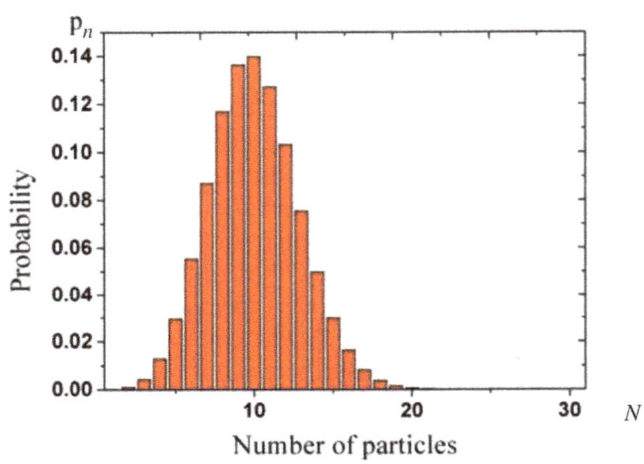

Figure 12.2. The distribution of the number of particles in volume V_1. The full volume V contains $N = 50$ particles.

to the ensemble gives rise to an increase of free energy $\langle E\rangle/\langle N\rangle$[1]. It follows that the change of energy due to adding *dn* particles into the same volume may be written in a differential form

$$dE = T\,dS + \mu\,dn, \qquad (12.2)$$

where the value μ is called the *chemical potential*. This name originates from chemical reaction physics, where non-conservation of particles is a typical phenomenon. Equation (12.2) means that the chemical potential is the energy cost of adding a particle at constant entropy.

Equation (12.2) modifies the Gibbs distribution of equation (11.6). Now the ensemble assigns a probability to each distinct microstate given by the following exponential,

[1] More rigorously it follows from Euler's theorem for a homogeneous function.

$$f_N(E) = \frac{1}{Q_N} e^{\frac{N\mu - E}{kT}}, \tag{12.3}$$

which is inherent to a *grand canonical ensemble*. In analogy to equation (9.12) and equation (11.1) this distribution generates a grand partition function Q_N and grand free energy $F(N, V, T)$, which is particulary useful for the description of processes occurring in open systems where particle number can vary but T and μ remain fixed.

Problem 12.1. The system of energy E contains n_1, n_2, ... molecules of different type. Prove that $T\,dS = dE - \sum \mu_i\,dn_i$.

Problem 12.2. Let $N = 10^4$ free particles of $m = 1.0$ obeying the Maxwell distribution of momenta fill the volume $V = 10 \times 10$ with PBCs. Using 2D simulation find the fluctuation of the number of particles and the energy in one part of the volume $V/4$ (see figure 12.2) as a function of the temperatures in the interval $kT = 0.1$–10.

Further reading

Pathria R K 1996 *Statistical Mechanics* (Oxford: Butterworth-Heinemann)

IOP Publishing

Principles of Statistical Physics and Numerical Modeling

Valeriy A Ryabov

Chapter 13

Isobaric ensembles

Up to now we have assumed that the volume of the system is constant. This requirement could be provided by something other than the boundary conditions for coordinates. The external force F introduced along the direction of the radius (see figure 7.3) and equal to the radial force of equation (7.4), $F_R = F$, leads to equilibrium where the length of the chain remains unchanged. Similarly, in an isotropic 3D case the volume does not vary if the internal pressure P of equation (7.8) is equal to the external one $P = P^{\text{ext}}$. Otherwise, if a volume is allowed to change, the difference $P - P^{\text{ext}}$ initiates dynamics for the volume.

According to equation (7.9), the increment of energy caused by the action of a constant external pressure is given by the term $\Delta U = P^{\text{ext}}V$. The sign of this expression takes into account the opposite direction of the external pressure compared to the internal pressure. Hence the Lagrangian of this joint dynamics takes the form

$$\mathscr{L}_P = \mathscr{L} - P^{\text{ext}}V, \tag{13.1}$$

where $\mathscr{L}$ refers to pressure equal to zero.

Consider first a 1D case that is a full analog of a system with uniaxial deformation. As we saw in chapter 7, this kind of deformation can be introduced by a radius in a polar coordinate system. Taking into account the expression $\mathscr{L}$ of equation (7.2), the Lagrangian of equation (13.1) becomes

$$\mathscr{L}_P = \mathscr{L} - P^{\text{ext}}L, \tag{13.2}$$

where $L = L_0R/R_0$. The new equations of motion following from this according to equation (1.3) take the form

$$\begin{aligned} \dot{x}_i &= R\dot{\theta}_i + \dot{R}\theta_i, \\ m(R^2\dot{\theta}_i)^{\cdot} &= RF_i, \\ Nm\ddot{R} &= Nf_R - P^{ext}\frac{L_0}{R_0}, \end{aligned} \tag{13.3}$$

where the radial force f_R is given by equation (7.3). Returning to the Cartesian coordinates $x_i = R\theta_i$, velocities $v_i = R\dot{\theta}_i$ and stretch ratio $\lambda = R/R_0$, and using the notion of the 1D internal pressure P of equation (7.5), rewrite these equations in the form

$$\begin{aligned}\dot{x}_i &= v_i + \frac{\dot{\lambda}}{\lambda}x_i,\\ \dot{v}_i &= \frac{F_i}{m} - \frac{\dot{\lambda}}{\lambda}v_i,\\ M_B\ddot{\lambda} &= f^{\lambda} - P^{\text{ext}}d_0,\end{aligned} \tag{13.4}$$

where $x_i = \lambda x_{i0}$ and $M_B = mR_0^2$ is the so-called *barostat mass*. It affects the ratio between the frequencies of thermal vibrations of atoms and the vibrations of the length of the chain. The choice of a specific magnitude M_B determines the time of approaching of equilibrium. At $M_B \gg m$ the equilibrium is reached too slowly, at $M_B \ll m$ it may not be reached at all. This question is discussed in appendix A.6. Nevertheless, the choice of the magnitude of M_B is not so important since, in equilibrium, the energy per this additional degree of freedom is $\sim kT$ regardless of the magnitude of the barostat mass. Therefore, when approaching equilibrium the contribution of a length vibration to the full energy becomes negligible $\tilde{}1/N$.

The equations of motion of (13.4) conserve the *instantaneous enthalpy*

$$\Phi = \frac{1}{2}\sum_{i=1}^{N}\frac{m\,v_i^2}{2} + \frac{NM_B}{2}\dot{\lambda}^2 + U + P^{\text{ext}}L. \tag{13.5}$$

Therefore, this ensemble is called *isobaric–isoenthalpic*.

Figure 13.1 shows the change of the chain period $d = L/N$, the internal pressure P, the mean kinetic energy of the particle K/N (~instantaneous temperature) and the barostat energy $K_V = \frac{M_B}{2}\dot{\lambda}^2$, calculated according to the equations of motion of (13.3) with an external pressure $P = 0.3$. The initial values of dynamical variables are $\dot{\lambda}(0) = 0$, $\lambda(0) = d_0 = 1.0$. All model parameters for a chain of atoms, including the potential U, are the same, as in the examples in chapters 6 and 7 (details of the calculation are given in appendix A.6).

The equations of motion of (13.4) can be easily extended to 2D and 3D systems under the uniaxial pressure P_α^{ext} of equation (7.7). In this case, the joint dynamics of uniaxial deformations relates only to three axes. The animation in figure 7.4 shows the solution of these equations in an ideal 2D lattice of atoms. The elongation is introduced along one direction, arbitrarily oriented (between the two yellow points) in the plane, while the length of the cell along another perpendicular direction remains unchanged. The top graph shows the lattice on the surface of a cylinder of varying radius. The bottom graph shows a scan cylinder, e.g. the real coordinates x, y. Details of calculations are given in appendix A.8, which also provides simulations for the 3D case.

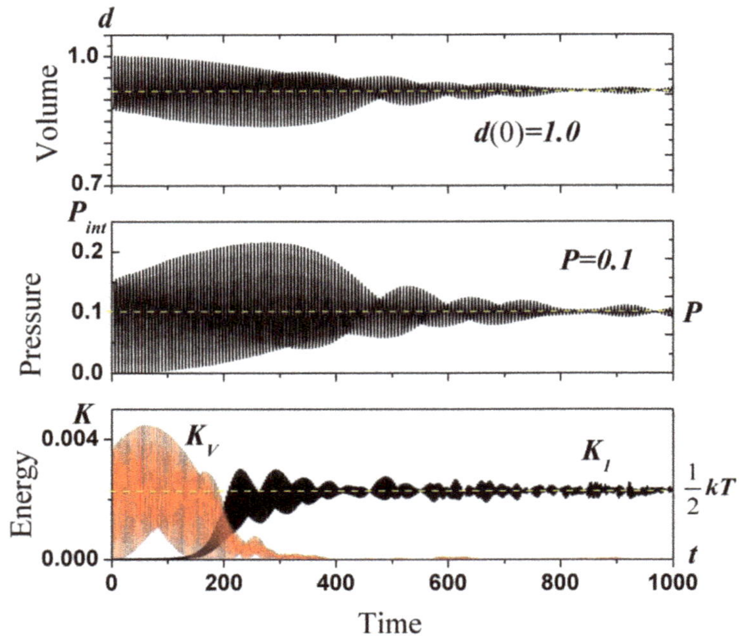

Figure 13.1. The dynamics of an isoenthalpic ensemble of a chain of interacting particles. The equilibrium values are shown by dashed lines.

It follows from calculations that near the equilibrium, the internal pressure tends to external, and oscillation energy of the barostat tends to zero as $K_V\tilde{~}kT/N$, as it must be due to equipartition theorem. Thus in the thermodynamic limit $N \to \infty$ the value

$$\Phi = E + P^{\text{ext}}V, \tag{13.6}$$

called *enthalpy* is a generalization of the definition of the energy in the case of a constant pressure and varying volume. In equilibrium, $\partial\Phi/\partial V = 0$ the volume of the system does not change and internal pressure is counterbalanced by external pressure:

$$\langle P \rangle = P^{\text{ext}}. \tag{13.7}$$

As is evident from the mechanical derivation, this statement needs the proof of another thermodynamical ensemble, the *isothermal–isobaric ensemble*. This ensemble merges the canonical and isobaric ensembles. From a formal point of view, it can be explored starting from a microcanonical description of a system coupled to both a thermal reservoir and a mechanical piston. The joint Hamiltonian includes the independent additional dynamical variables R of equation (13.2), responsible for deformation, and γ of equation (10.5), responsible for maintaining the temperature. Leaving aside, for now, the correspondent mechanistic derivation (it will be carried out in chapter 17), consider the proof of equality equation (13.7) in terms of this ensemble.

The inclusion of the additional dynamic variable λ into the canonical distribution of equation (9.10) obeys the same rules as any degrees of freedom of the particle in phase space $\tilde{f}(\mathbf{x}, \mathbf{p}, \lambda, p_\lambda)e^{-\beta\Phi}$. Therefore, after integration over the barostat momentum p_λ, we have

$$\tilde{f}(\mathbf{x}, \mathbf{p}, \lambda) = \frac{1}{\widehat{Q}} \exp(-\beta[H_\lambda(\mathbf{x}, \mathbf{p}) + P^{\text{ext}} V(\lambda)]). \tag{13.8}$$

Here $V = V_0\lambda^3$ and $\widehat{Q}$ is a normalization constant.

$$\widehat{Q} = \int_0^\infty d\lambda e^{-\beta P^{\text{ext}} V} Q(\lambda, T) \tag{13.9}$$

with the normalization factor of equation (9.9)

$$Q(\lambda, T) = \int d\mathbf{x}\, d\mathbf{p}\, e^{-\beta H_\lambda(\mathbf{x}, \mathbf{p})} \tag{13.10}$$

related to a canonical ensemble with the free energy $F(\lambda, T)$.

The average of the energy over the distribution of equation (13.8), by analogy with equation (11.2) leads to

$$\langle \Phi \rangle = \langle H_\lambda \rangle + P^{\text{ext}} V = E + P^{\text{ext}} V = -\frac{\partial}{\partial \beta} \ln \widehat{Q}. \tag{13.11}$$

The average of the internal pressure yields

$$\begin{aligned}\langle P \rangle = -\left\langle \frac{\partial H}{\partial V} \right\rangle &= -\frac{1}{\widehat{Q}} \int_0^\infty d\lambda e^{-\beta P^{\text{ext}} V} \int d\mathbf{x} d\mathbf{p} \frac{\partial H_\lambda}{\partial V} e^{-\beta H_\lambda(\mathbf{x}, \mathbf{p})} = \\ &= \frac{1}{\widehat{Q}} \int_0^\infty d\lambda e^{-\beta P^{\text{ext}} V} \int d\mathbf{x} d\mathbf{p} kT \frac{\partial e^{-\beta H_\lambda(\mathbf{x}, \mathbf{p})}}{\partial V} = \\ &= \frac{1}{\widehat{Q}} \int_0^\infty d\lambda e^{-\beta P^{\text{ext}} V} kT \frac{\partial F(\lambda, T)}{\partial V}.\end{aligned} \tag{13.12}$$

Performing integration by parts, we have

$$\begin{aligned}\langle P \rangle &= \frac{1}{\widehat{Q}} [e^{-\beta P^{\text{ext}} V} k\, T\, F(\lambda, T)] \Big|_0^\infty - \frac{1}{\widehat{Q}} \int_0^\infty dV kT \frac{\partial e^{-\beta P^{\text{ext}} V}}{\partial V} F(\lambda, T) = \\ &= P^{\text{ext}} \frac{1}{\widehat{Q}} \int_0^\infty d\lambda e^{-\beta P^{\text{ext}} V} F(\lambda, T) = P^{\text{ext}}.\end{aligned} \tag{13.13}$$

Although this average refers to a constant predominant temperature, we omitted index T everywhere under the parenthesis, since the same relation stands for an isobaric–isoenthalpic ensemble in which microcanonical temperature follows from initial conditions. Note that in the derivation above we could have used the volume as an independent variable. In this case, changing the variable $\lambda \to V$ reduces to redefining distribution and normalization constant (see problem 13.1).

Finally, another isobaric ensemble can be constructed if added up to the isothermal–isobaric ensemble variable number of the particles (see the previous chapter). Repeating calculations similar to those carried out in chapter 12, the canonical distribution function of an ensemble with a variable number of particles can be written as

$$\tilde{f}(\mathbf{x}, \mathbf{p}, V, n) = \frac{1}{\widehat{Q}_n} \exp\left(-\beta[H_n(\mathbf{x}, \mathbf{p}) + PV + \mu n]\right). \tag{13.14}$$

In addition to the phase space, this distribution also drives the additional variables V and N, forming together *the grand canonical ensemble.*

Problem 13.1 Prove the *work virial theorem* using the volume as an independent variable: $\langle PV \rangle + kT = P^{\text{ext}}\langle V \rangle$.

Further reading

Tuckerman M E 2010 *Statistical Mechanics: Theory and Molecular Simulation* (New York: Oxford University Press)

IOP Publishing

Principles of Statistical Physics and Numerical Modeling

Valeriy A Ryabov

Chapter 14

Thermodynamic potentials

In a purely mechanical system, energy can be stored in the form of potential energy and subsequently retrieved. The same is true for thermodynamic systems. Energy can be stored in a thermodynamic system by doing work on it or by increasing the temperature and eventually that energy can be retrieved in the form of work. We have already seen that in a system with random microstates, this performs the role of free energy. However, there are as many different forms of free energy in a thermodynamic system as there are combinations of macroscopical *controlled parameters* V, λ, P, T. Unlike *microstates*, a set of these values, the *macrostates* also fully define the system, but from the thermodynamic point of view. In this chapter, we shall discuss the four possible thermodynamical functions which can be built on the basis of these macrostates (uniaxial deformation can be considered quite analogously to volume deformations).

The total differential of the energy of equation (11.12) is

$$dE = TdS - PdV. \tag{14.1}$$

The enthalpy can be found from this equation

$$d\Phi = dE + PdV = TdS + VdP. \tag{14.2}$$

The sign of the average and the index of a constant number of particles is omitted here for convenience.

For the free energy, we had equation (11.13)

$$dF = -SdT - PdV. \tag{14.3}$$

To complete the possible opportunities to represent all thermodynamic values through the partial derivatives with respect to variables V, P and T, it is necessary to define a total differential in variables P and T. For this purpose, we note that $PdV = d(PV) - VdP$, so after substitution of equation (14.3), one obtains

doi:10.1088/978-0-7503-1341-4ch14

$$dG = -SdT + VdP. \tag{14.4}$$

These formulas allow the determination of all four quantities[1], the energy E (*internal energy*), the free energy F (*the Helmholtz free energy*), the enthalpy Φ and *the Gibbs free energy* G, by means of one of them, accounting for partial derivatives. The thermodynamics, operating with these four quantities by means of differential relations, becomes a closed axiomatic theory, which does not require knowledge of these potentials at the microscopic level to obtain the results.

These quantities play a role analogous to the potential energy of a mechanical system, and for that reason, they are also called the *thermodynamic potentials.*

The differential relations above allow changing thermodynamic variables as the control variables. The most convenient way to do this is *Legendre's conversion.* It allows one to express a function through its derivative and provides a means by which one can determine how the energy functions for different sets of thermodynamic variables are related. An example is given below for functions of a single variable.

Consider a function $f(x)$ and its derivative $y' = f(x)$. The equation $y = g(x)$ defines a variable transformation from x to y. Figure 14.1 shows a point at coordinate x_0 with the tangent line at $f(x_0)$ drawn in red. If we specify the slope of this line as $f'(x_0)$, then $f(x_0) = f'(x_0)x_0 + a(x_0)$. This relation must hold for any x:

$$f(x) = f'(x)x + a(x). \tag{14.5}$$

Introducing the functional inverse of $g(x)$: $x(y) = g^{-1}(y)$ and solving the same for $a(x) = a(g^{-1}(y))$ gives

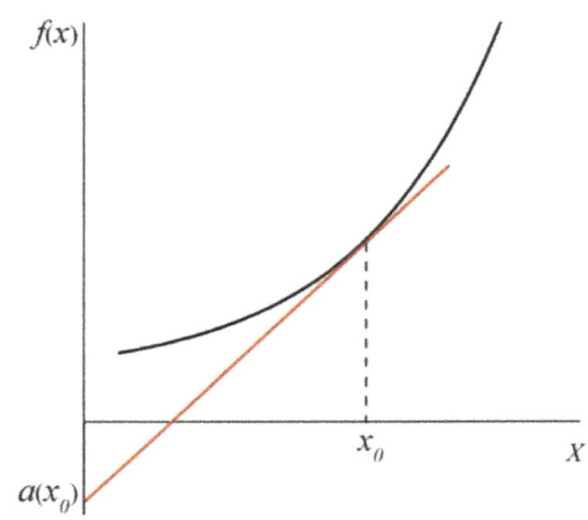

Figure 14.1. To Legendre's conversion.

[1] To these should be added a fifth quantity, the *Grand potential*, which deals with ensembles of variable numbers of particles. It is obtained from the considered potentials by adding to their differentials the term $\mu\ dN$ just as in equation (12.2). This means that the chemical potential can be obtained by differentiation of any quantity E, F, G, Φ on N, although expressed through different variables.

$$a(x) = f(g^{-1}(y)) - yg^{-1}(y)) \equiv \phi(y), \tag{14.6}$$

where $\phi(y)$ is known as the Legendre transform of $f(x)$. In shorthand notation, one writes

$$\phi(y) = f(x) - xy. \tag{14.7}$$

How does a Legendre transform work? In the microcanonical ensemble, the entropy S is a function V and E, i.e., $S = S(V, E)$. This can be inverted to give the energy as a function of V and S, i.e., $E = E(V, S)$. Let us use this in the equation $T = \partial E/\partial S$ of (11.10) to change the variable from S to T. Construct a set of equivalence $f = E$, $x = S$, $y' = \partial E/\partial S = T$. Then the Legendre transform gives

$$\phi(V, T) = E(V, S) - S\frac{\partial E}{\partial S}. \tag{14.8}$$

The quantity $\phi(V, T)$ being the energy for the constant volume and entropy is nothing but the free energy. Hence we come to equation (11.8)

$$F = E - TS. \tag{14.9}$$

This procedure can be extended to the case of two variables to transform the thermodynamic relations obtained above.

Problem 14.1 In equation (14.1) change the variable from V to P.

Problem 14.2. Prove the cyclic rule: $(\frac{\partial T}{\partial P})_\Phi(\frac{\partial \Phi}{\partial T})_P(\frac{\partial P}{\partial \Phi})_T = -1$.

Further reading

Reichl L E 1998 *A Modern Course in Statistical Physics.* (New York: Wiley)

IOP Publishing

Principles of Statistical Physics and Numerical Modeling

Valeriy A Ryabov

Chapter 15

Finite ensembles and thermodynamics

We saw that partition functions related to microcanonical and canonical ensembles are different. How large is this difference for the free energy? At first, to clear up this question, express the density of states $\Omega(E)$ of equation (8.5) through the partition function $Q(\beta)$ of equation (9.9) with the help of the Laplace transformation. For any function, it is given by $F(\lambda) = \int_0^\infty e^{-\lambda x} f(x) dx$. The transformation of $\Omega(E)$ of equation (8.5) with respect to E yields

$$Q(\lambda) = \int_0^\infty dE e^{-\lambda E} \int \delta(E - \mathscr{H}(\mathbf{x}, \mathbf{p})) d\mathbf{x} d\mathbf{p}. \tag{15.1}$$

Carrying out the integration over E we obtain

$$Q(\lambda) = \int_0^\infty dE e^{-\lambda E} \int \delta(E - \mathscr{H}(\mathbf{x}, \mathbf{p})) d\mathbf{x} d\mathbf{p} = \int_0^\infty dE e^{-\lambda E} \Omega(E). \tag{15.2}$$

By identifying $\lambda = \beta$ we come to the canonical partition function of equation (9.9).

Now trace the difference between the microcanonical and canonical ensembles on an example of the definition of the temperature within the microcanonical and canonical ensembles. It turns out that this difference in a thermodynamic limit ($N \to \infty$) extends to other important macroscopic averages. Let us illustrate this with the example of differences in the expressions for the heat capacity of equations (6.6) and (11.11). For convenience, we introduce specific values of energy $e = E/N$, the entropy of equation (11.10) $s_N = S_N/N$ and the free energy $f_N = F_N/N$ per particle relating to the canonical ensemble. Using the energy representation for a partition function of equation (9.11) and taking into account equation (11.8), we have

$$Q = e^{-\beta N f_N} = \sum_e e^{-\beta N[e - s_N(e)]}. \tag{15.3}$$

In the thermodynamic limit $N \to \infty$, this sum can be treated as the Laplace transform of equation (15.2) with respect to the energy e. It satisfies the relation

$$f_\infty = \min_e(e - Ts_\infty). \tag{15.4}$$

Hence the partition function via a Laplace transform becomes available in the thermodynamic limit. It enables us to trace a connection between all values relating both to the finite system s_N, f_N and to their thermodynamic limit s_∞, f_∞.

Microcanonical energy can be determined by the solution of equation (11.10): $1/T_N^{\text{mic}} = \partial s_N/\partial e$ through the Legendre's conversion of equation (14.7) to yield the microcanonical equation-of-state $e_N^{\text{mic}} = e_N^{\text{mic}}(T_N^{\text{mic}})$. The differentiation of this with respect to T_N^{mic} gives

$$c_N^{\text{mic}}(T_N^{\text{mic}}) = \frac{\partial e_N^{\text{mic}}}{\partial T_N^{\text{mic}}} = -\left(\frac{\partial s_N}{\partial e}\right)\left(\frac{\partial^2 s_N}{\partial e^2}\right)^{-1}. \tag{15.5}$$

For a canonical ensemble of equation (10.8), we obtain

$$c_N^{\text{can}}(T) = -T\frac{\partial f_N}{\partial T^2} = N\frac{\langle e^2\rangle_N - \langle e\rangle_N^2}{kT^2}, \tag{15.6}$$

For sufficiently large N, the difference between the corresponding values in all three cases becomes negligible $\sim 1/\sqrt{N}$. Note that the ratio of equation (15.6) represents an example of the use of so-called *thermodynamic fluctuations* taking place even in a limit $N \to \infty$.

Let us illustrate the difference between ensembles and their thermodynamical limit by calculating the heat capacity of equations (15.5) and (11.4) within a model which is of great importance in statistical physics, the Ising model. Instead of vibrating atoms, place in cells of a crystal lattice magnet moments—spins interacting with each other and having only two possible values $s_i = \pm 1$. The energy of such a system is

$$E = -J\sum_{i,j} s_i s_j. \tag{15.7}$$

Let us assume a constant $J > 0$, and consider interaction only between nearest neighbors. Discretization of states allows not only a simple numerical calculation of all mean values, but also obtaining their analytical expressions for the 1D and 2D cases.

According to the Boltzmann distribution of equation (9.14), the probability of any configuration of spins $\alpha = (s_1, s_2, \ldots, s_N)$ is

$$P_\alpha = \frac{1}{Q}\exp\left(-\frac{E_\alpha}{kT}\right), \tag{15.8}$$

where the partition function

$$Q = \sum_\alpha P_\alpha. \tag{15.9}$$

Despite the simplicity of this expression, its direct calculation is problematic even with current computational power. The number of states of the system, although still of course finite, nevertheless is very great. For example, in a square 2D lattice with $N = L \times L$, their number is equal to $2^{N \times N}$. Therefore, to calculate the free energy of the system by the formulas of equations (15.9) and (11.1), we will use the *Metropolis algorithm* which allows moving along the most probable configurations when calculationg the sums of an equation such as (15.9), providing thereby the smallest error in the determination of the partition function at a manageable number of steps. This algorithm is based on application of the *principle of detailed balance*

$$P(s_i)P(s_i \rightarrow s_j) = P(s_j)P(s_j \rightarrow s_i). \tag{15.10}$$

This principle means that the probabilities of forward and backward processes of a state change are equal. In particular, it implies that around any closed cycle of states the probability of $s_i \rightarrow s_j \rightarrow s_k \rightarrow s_i$ is the same as for the inversed cycle $s_i \rightarrow s_k \rightarrow s_j \rightarrow s_i$. This can be proved by substitution from the definition of equation (15.10).

From equation (15.10) the probability of change of a spin $P(s_i \rightarrow s_j)$ is defined by

$$\frac{P(s_i \rightarrow s_j)}{P(s_j \rightarrow s_i)} = \frac{P(s_j)}{P(s_i)} = \exp\left(-\frac{E_j - E_i}{kT}\right). \tag{15.11}$$

Choosing in a random way the turn of a spin on arbitrary site i with probability

$$w_{ij} = \begin{cases} \exp\left(\dfrac{E_i - E_j}{kT}\right), & E_i < E_j; \\ 1, \quad E_i > E_j, \end{cases} \tag{15.12}$$

one can calculate the partition function in the process of simulations as an average of the values P_α. The main feature of such an algorithm, related to so-called Markov algorithms (which do not take into account the history of a process of transitions) is a guaranteed trend of the system to equilibrium (steady-state distribution), regardless of the initial configuration of spins.

The result of calculation of a thermal capacity $c(T)/k$ as a function of temperature for a square 32×32 lattice of spin is given in figure 15.1 (details are given in appendix A.5). A comparison was made with the known exact analytical result of Onsager of equation (15.4) for $J = 1$. This example illustrates a degree of convergence of results of both types of ensembles to an exact result. It is seen that the largest error caused by a small number of atoms is observed in the region close to $T = T_c = 2.23$ (the critical point of phase transition) where the heat capacity has a discontinuity. The reason for this phase transition will be clear in chapter 22. However, away from the singularity point we observe their full coincidence.

There are also other opportunities for the determination of $Q(E)$ and the ensemble averages by the spin dynamics in the framework of the microcanonical

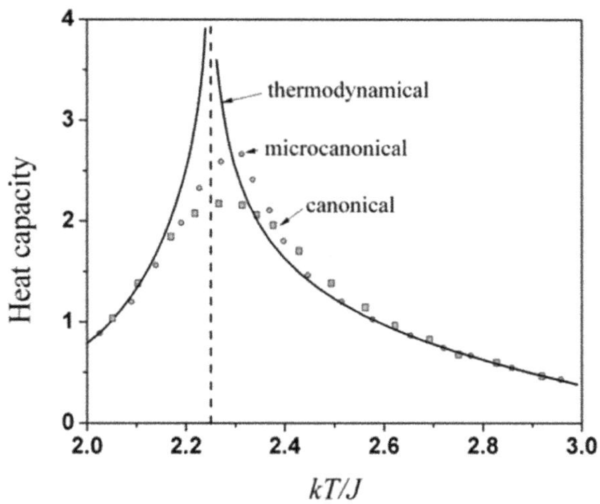

Figure 15.1. The dependence of specific heat on temperature. The solid line shows the exact solution—the thermodynamic limit. ○ = calculations for the microcanonical ensemble. □ = calculations for the canonical ensemble.

ensemble (at the fixed energy E). In particular, such calculations are possible without a given temperature, determining it in the course of calculation, or at fixed total spin and a predetermined temperature.

Further reading

Krauth W 2006 *Statistical Mechanics: Algorithms and Computations* (New York: Oxford University Press)

Rugh H H 1997 *Phys. Rev. Lett.* **78** 772

IOP Publishing

Principles of Statistical Physics and Numerical Modeling

Valeriy A Ryabov

Chapter 16

Equation-of-state for a non-ideal gas

We have already encountered the equation-of-state when we obtained the relationship between temperature, pressure, volume and the number of particles in a system. In the thermodynamic limit, these relations do not depend on the kind of ensemble used for calculations. One of the simplest equations-of-state is the ideal gas law of equation (7.10)

$$P = \rho kT, \tag{16.1}$$

where $\rho = N/V$ is the density. However, this equation becomes increasingly inaccurate at higher pressures or lower temperatures, when an interaction between particles becomes noticeable, causing gas–liquid–solid transitions. So the equation-of-state must include other important features of all substances relating to the spatial distribution of particles.

It is convenient to derive an equation-of-state for a gas with interaction, a *non-ideal gas*, within the isothermal ensemble developed in chapter 10. Let us calculate isotherms for 2D hexagonal lattice (see figure A.8 in appendix A.4). The initial positions of $N = 900$ atoms with the pair potential of interaction of equation (1.14) were set in sites of the hexagonal lattice with the period $d_0 = 1.1$–2.5. Initial momenta were given in the form of a uniform distribution in the interval $|p_{i\alpha}| < \delta$, $\alpha = x, y$ with values $\delta = 0.5$. After a small period to establish equilibrium, the kinetic energy matches the external values $\overline{K_i} = kT$ and the internal pressure is fixed. The results of the calculations are shown in figure 16.1. During a run, the equilibrium values of enthalpy $\Phi(T, V) = E + \overline{P}_{\text{int}}V$ were also recorded. They will be needed in chapter 18 below.

The initial portion of all the curves presents a linear dependence inherent in an ideal 2D gas: $P = \rho kT$. In general, by means of a fitting procedure, these curves can be approximated by the equation

$$P = \frac{\rho kT}{1 - b\rho} - a\rho^2 \tag{16.2}$$

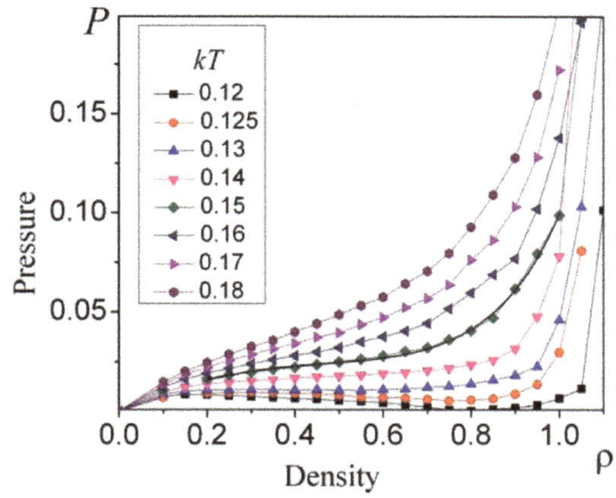

Figure 16.1. The equation-of-state of a real gas. The solid red line on the isotherm $kT = 0.15$ is the van der Waals equation with parameters $b = 0.667$, $a = 0.226$ and $kT = 0.11$.

with constants $a \approx 0.2$ and $b = 0.67$, known as the *van der Waals equation* (the solid line in the top part of figure 16.1). At low density, this equation describes an ideal gas, at a large density with $b\rho \sim 1$ it is a liquid with limited compressibility (a large slope of the curves in the top right of the graph). We will return to the distinction of these areas depending on the density of atoms in chapter 21.

The origin of the denominator in equation (16.2) can be interpreted as a reduction of the effective volume of the ensemble due to a change of the size of particles. An ideal gas consists of point particles, in realilty they have some radius due to repulsive interaction, which should be excluded from the volume of the system: $V \to V - Nb$. The second term in equation (16.2) follows from the virial contribution in the expression for the pressure of equation (7.8).

To obtain the magnitude of these constants, add and subtract unity from the integrand of free energy of equation (11.15), so that

$$F = F_{\text{ид}} - kT \ln \left\{ 1 + \frac{1}{V^N} \int (e^{-\beta U(x)} - 1) dx \right\}. \tag{16.3}$$

Here we use that $\int dx = V^N$. In a gas, atoms are far apart, so the integrand is non-zero when two atoms are close to each other. Assuming that this couple can only be a single at any moment at points $\boldsymbol{r}_1$ and $\boldsymbol{r}_2$, it can be chosen in $N(N-1)/2$ ways. The result of integration over the remaining coordinates in the integral of equation (16.3) yields

$$\frac{N(N-1)}{2} V^{N-2} \int (e^{-\beta U(\boldsymbol{r}_{12})} - 1) d\boldsymbol{r}_1 d\boldsymbol{r}_2, \tag{16.4}$$

where $\boldsymbol{r}_{12} = \boldsymbol{r}_2 - \boldsymbol{r}_1$. Now make here the replacement of variables $\boldsymbol{r}_1, \boldsymbol{r}_2 \to \boldsymbol{r}_{12}, \boldsymbol{R}$, so that the integration on the coordinate of the center of gravity $\boldsymbol{R}$ in this integral gives V. Further, take into account that a small term is added to the unit in a logarithm of equation (16.3). Hence using the expansion $\ln(1 + y) \approx y$, the free energy of non-ideal gas can be written in the form

$$F = F_{id} + \frac{N^2 B(T)}{V}, \tag{16.5}$$

where

$$B(T) = \frac{1}{2}\int (e^{-\beta U(r_{12})} - 1)d\boldsymbol{r}_{12}. \tag{16.6}$$

Now take into account that for small distances $r \ll 1$ the pair potential $U(r) \gg 1$, while at large ones: $U(r) \ll 1$. Then, separating the integration area in equation (16.6) into two parts: $r_{12} < r_0$ where $e^{-\beta U(r_{12})} \to 0$, and $r_{12} > r_0$ where $e^{-\beta U(r_{12})} \to 1 - \beta U(r_{12})$, we obtain

$$B(T) = b - \frac{a}{T}. \tag{16.7}$$

Substituting this into equation (16.5) we obtain

$$F = F_{id} + \frac{N^2}{V}(Tb - a). \tag{16.8}$$

Typically, $Nb \ll V$, so one can present the product $N^2 T b \approx N T \ln\left(1 - \frac{Nb}{V}\right)$ to write the free energy in the form

$$F = F_{id} - N T \ln\left(1 - \frac{N b}{V}\right) - \frac{N^2 a}{V}. \tag{16.9}$$

Differentiation of this expression on the volume $P = -\partial F/\partial V$ leads to the van der Waals equation of (16.2).

The entropy follows from equation (16.9)

$$S = S_{id} + N \ln\left(1 - \frac{Nb}{V}\right), \tag{16.10}$$

and the energy $E = F + TS$

$$E = E_{id} - \frac{N^2 a}{V}. \tag{16.11}$$

Direct calculation of the constant $B(T)$ in equation (16.7) for the 2D model above gives $a = 0.12$ and $b = 0.6$, close to the result of the fitting of equation (16.2). For lower temperatures and high atomic density, the state of the ensemble is no longer that of a gas, but another *aggregate* state. This will be discussed in more detail in chapter 21. Here we only note that in the condensed phase the equation-of-state is expressed by the more general *constitutive equation* of (13.7): $\langle P \rangle = P^{\text{ext}}$.

Problem 16.1 Find the difference $C_p - C_V$.

Further reading

Benenson W, Harris J W, Stocker H and Lutz H 2001 *Handbook of Physics* (New York: Springer)

IOP Publishing

Principles of Statistical Physics and Numerical Modeling

Valeriy A Ryabov

Chapter 17

Processes in an ideal gas

The equation-of-state relates to an equilibrium macrostate in which the temperature, pressure and volume cannot change spontaneously, since this would increase the thermodynamical potential. Transitions between them become possible if the system is exposed to an external action such as the heating/cooling or compression/expansion. Herewith a time evolution of the control parameters of the system T, P and V passes through a sequence of intermediate *quasistatic states* to a final state. If their change occurs rather slowly, so that in each intermediate state the equilibrium has time to be established, then in this *thermodynamic process*, we can use differential relations for the thermodynamic potential derived in chapter 14 (see table 17.1). They give the answer to the important question of how the thermal energy of the system transforms to work and vice versa.

Basically, as follows from the equation-of-state, three possible processes to be kept constant are the volume $\Delta V = 0$ (an *isochoric process*), the pressure $\Delta P = 0$ (an *isobaric process*) or the temperature $\Delta T = 0$ (an *isothermal process*). If these three processes occur in the closed system (interacting with the environment) they require a heat supply and work to be done.

Heating is a process of the adding of the energy ΔQ to a system because its temperature differs from the temperature of the thermostat. In general, this process is accompanied by performing work dW, therefore the change of energy is given by

$$dE = dQ + dW. \tag{17.1}$$

The pressure–volume work is defined as:

$$W = -\int_{V_1}^{V_2} PdV. \tag{17.2}$$

Here we used the sign convention for work, where positive work is work done on the system.

doi:10.1088/978-0-7503-1341-4ch17

Table 17.1. Processes in an ideal gas.

Process	Isobaric $\Delta P = 0$	Isochoric $\Delta V = 0$	Isothermal $\Delta T = 0$	Adiabatic $\Delta Q = 0$
Differential	$dE = dQ - PdV$ $d\Phi = dQ + VdP = dQ$	$dE = dQ$	$dE = 0$ $dQ = -dW$	$dE = PdV$ $d\Phi = VdP$
Work $W = -\int_{V_1}^{V_2} PdV$	$-P(V_2 - V_1)$	0	$NkT \ln \frac{V_2}{V_1}$	$C_V(T_2 - T_1)$
Energy $\Delta E = Q + W$	$Q + \mathrm{W}$	Q $C_V(T_2 - T_1)$	0 $Q = -W$	W $C_V(T_2 - T_1)$
Enthalpy $\Delta\Phi = \Delta(E + PV)$	$C_P(T_2 - T_1)$	$Q + V\Delta P$	0	$C_P(T_2 - T_1)$
Entropy $\Delta S = N\int_{T_1}^{T_2} \frac{C}{T} dT$	$C_P \ln \frac{T_2}{T_1}$	$C_V \ln \frac{T_2}{T_1}$	$Nk \ln \frac{V_2}{V_1}$	0

The change in the entropy of equation (8.15) in this process is

$$\Delta S = \int_{E_1}^{E_2} \frac{dE}{T(E)}. \tag{17.3}$$

We shall begin by discussing the question of how heating becomes work by using a fully mechanistic model of the isobaric ensemble schematically shown in figure 17.1. Calculations will be made for the ideal 2D gas with only a small modification of this model. Let the red moving cap of heavy mass M_B be balanced by the same lever weight, the barostatic mass. So only the green weights of mass $m_P \ll M_B$ produce the applied constant pressure $P^{\text{ext}} = m_P g/l$, where l is the transverse size of the cap. Let the velocity of atoms in a camera obey the Maxwell distribution with the variance $\langle mv_x^2\rangle = \langle mv_y^2\rangle = kT$. In each collision an atom with a velocity v_y, the momentum $m\ \Delta v_y$ transfers to the cap so that the total amount $\Delta p = 2\sum m\ \Delta v_y$ gained for the time τ determines the internal pressure $P = \Delta p/\tau l$. The presence of a large barostatic mass is necessary to gain sufficient momentum transfer for the determination of a pressure for a time τ. In turn, this time must be sufficiently small for a velocity of the displacement of the cap $\Delta y/\tau$ small enough to relate the internal pressure to a definite temperature $T \gg \frac{\Delta T}{\Delta t}\tau$. In addition, an appropriate choice of barostatic mass can essentially depress oscillations of the cap arising as a result of dynamics. Correspondingly, the equation of the heavy cap is

$$M_B \ddot{y} = m_P g - Pl. \tag{17.4}$$

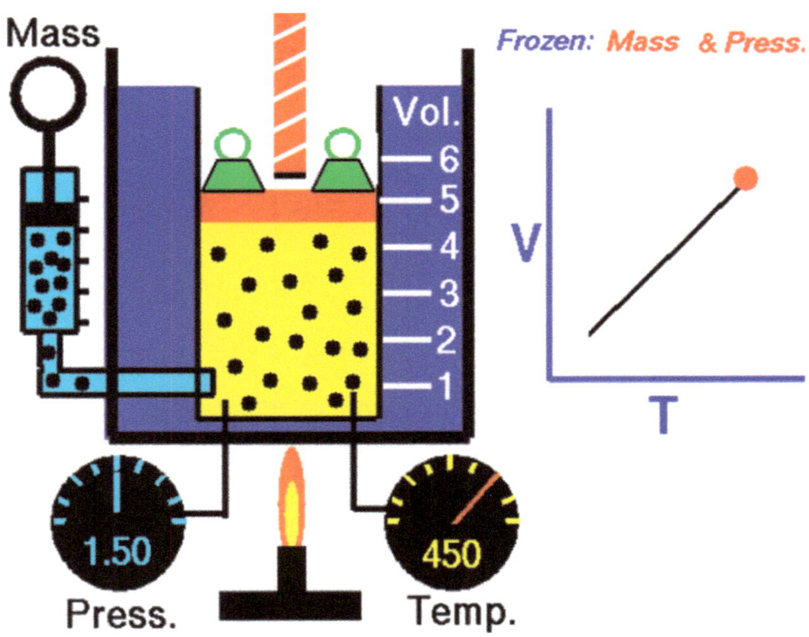

Figure 17.1. Performance of work in isobaric ensemble. Source: grc.nasa.gov NASA. Animation available at http://iopscience.iop.org/book/978-0-7503-1341-4.

In figure 17.2 are shown the results of dynamics of this prototype of a steam nano-engine with parameters $m = 1$, $M_B = 2 \cdot 10^5$, $m_P g = 100$, $\tau = 10^3$, $\Delta T/\Delta\tau = 2 \cdot 10^{-7}$, $N = 2000$. Details of the computation algorithm are given in appendix A.11.

As is seen from the results, the internal pressure matches the external pressure well, $P = P^{\mathrm{ext}}$, and a heavy cap is making the oscillation of all macroscopic values small. You can verify from the figure that $\Delta E/N = \Delta kT \approx 2$ and $W/N = -mg \cdot \Delta y \approx -2$.

To explain these results consider at first the changes of enthalpy in the isobaric process for a 3D ideal gas. Applying the ideal gas law we have $\Delta E = (3/2)Nk\Delta T$ and $W = -P\Delta V = Nk\Delta T$. Substitution of these two ratios into equation (17.1) produces

$$Q = \frac{5}{2}Nk\Delta T = C_p(T_2 - T_1), \tag{17.5}$$

where $C_P = (5/2)Nk$ is the heat at a constant pressure. In this process, only the volume changes, $V_2 = V_1T_2/T_1$. Therefore, the enthalpy $\Delta\Phi = \Delta Q + V\Delta P = \Delta Q$. Executing the integration (17.3) taking into account equation (6.4) we obtain

$$\Delta S = \int_{T_1}^{T_2} \frac{N}{2}k\frac{dT}{T} = \frac{3}{2}Nk \ln \frac{T_2}{T_1}. \tag{17.6}$$

Use these formulas for the results shown in figure 17.2. The heat for the 2D gas of equation (17.5) goes to $C_p = 2Nk$ so that $\Delta Q/N = 2k(T_2 - T_1) \approx 2(2.26 - 0.26) = 4.0$. The work done is $W/N = -P(V_2 - V_1)/N = -mg(y_2 - y_1)/N \approx 100 \cdot (50 - 10)/1000 = 2.0$. As is expected, the change of the kinetic energy due to the increase of temperature is equal to $\Delta E = 2.0 \cdot N = \Delta Q + W = (4.0 - 2.0)N$. One can also verify that the gas law of equation (7.11) is fulfilled for the points where the kinetic energy of the barostat vanishes.

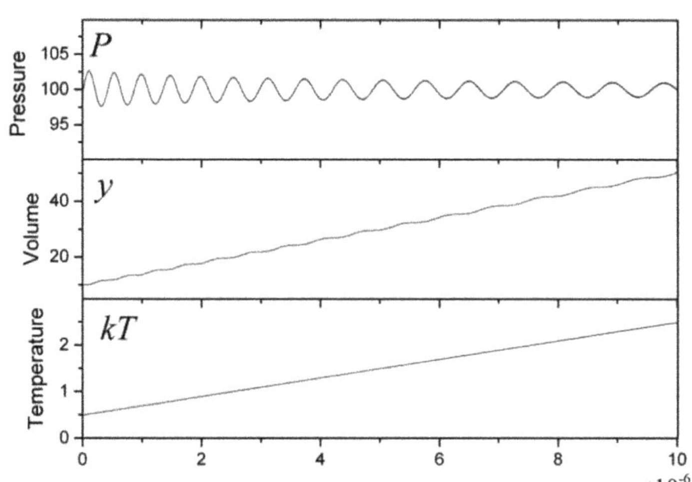

Figure 17.2. Isobaric heating.

In the *isochoric process*, the heating causes an increase in the temperature $T_1 \rightarrow T_2$ and pressure $P_1 \rightarrow P_2$,

$$\Delta Q = \int_{T_1}^{T_2} C_v dT. \tag{17.7}$$

A compression/expansion of gas at a constant temperature (*isothermal process*) leads to a change of the pressure according to $VP = \text{const} = NkT$. Thus, the gas expansion is propelled by absorption of heat energy

$$\Delta Q = -W = \frac{3}{2}\int_{V_1}^{V_2} \frac{NkT}{V} dV = \frac{3}{2} NkT \ln \frac{V_2}{V_1}. \tag{17.8}$$

In a thermally insulated system an *adiabatic process* occurs without heat transfer into or out of a system, so that $\Delta Q = 0$. The energy in the system is transferred to its surroundings only as work. In contrast to the processes considered above, we can use this approximation only in very rapid processes, in which there is not enough time for the transfer of energy as heat to take place into or out of the system. Another example of this process is the free expansion of a gas into a vacuum. The work done by the gas in the expansion is $W = PdV$ and the change in the internal energy $dE = C_V dT$. Therefore, from equation (11.12), it follows that

$$C_V dT = 0 - PdV = -PdV. \tag{17.9}$$

In a similar fashion, other results may be derived for processes on ideal gases. A summary of results is given in table 17.1. Consider in more detail the adiabatic cooling based on this differential equation. It occurs when the pressure on an adiabatically isolated system is decreased, allowing it to expand. The formula of equation (17.9) lends itself to a simple numerical evaluation in which the sequence of operations is the following. Starting from initial volume $V_1 = 1$, pressure $P_1 = 10$ and kinetic energy $E = \frac{3}{2}NkT_1 = 15$, decrease the volume with the small step $\Delta V = 0.003$. Deduct the work done at each step $\Delta W = P\Delta V$ from a current magnitude of the energy $E = E - \Delta W = \frac{3}{2}kT$. Determine a new value of the pressure $P = kT/V$ and enthalpy $\Phi = E + PV$. Then this step is repeated with the new values of V, P and T. The calculation finishes once $T = T_2$.

The result of this calculation is shown in figure 17.3. It is easy to verify that while the energy is represented by the line $E = \frac{3}{2}NkT$ the enthalpy behaves as $\Phi = \frac{5}{2}NkT$. Lying between two *isotherms* $kT_1 = 10$ and $kT_2 = 4.0$, the adiabatic curve can be fitted by the *polytropic process equation* $PV^{\frac{5}{3}} = 10$. As is clear from the sequence of operations, the area under this curve, the work, is equal to the change of energy (compare this with table 17.1).

In this process, an ideal gas undergoes a reversible (i.e. no entropy generation) adiabatic process. Of course, these results could have been reproduced by analytic calculations. However, the usefulness of numerical computations based on differential relations for thermodynamic potentials is in their simple extension to the more complicated equations-of-state for real substances.

Almost all the processes considered above can be reversed by inducing infinitesimal changes and, in so doing, leaving no change in either the system or surroundings. During these *reversible processes*, the entropy of the system does not increase and the system is in thermodynamic equilibrium with its surroundings.

It is possible to return to the initial state also by means of a *cyclic process* (figure 17.4). This consists of a linked sequence of processes considered above that involve the transfer of heat and work into and out of the system. Variation of volume, pressure and temperature eventually returns the system to its initial state so that $\Delta E = E_{\text{fin}} - E_{\text{in}} = 0$. The simplest is a circular cycle. It consists of two processes of expansion (1–2) and compression (2–1). Work is defined by the enclosed area, in the left-hand panel it is positive and in the right-hand panel it is negative. Circular processes underlie *heat engines*: the left-hand panel corresponds to the operation of a steam engine, and the right-hand panel to a refrigerator.

Another important example is the *Carnot cycle*, which acts as a heat engine. It consists of four steps (see figure 17.5), each of them calculated in the same manner as in figure 17.3. The first one (blue) is the reversible isothermal expansion of the gas at the 'hot' temperature T_1. During the next step (red) the gas is allowed to expand and it does work on the surroundings. The gas expansion is accompanied by absorption of heat energy Q of the equation (17.8) reservoir and results in an increase of entropy

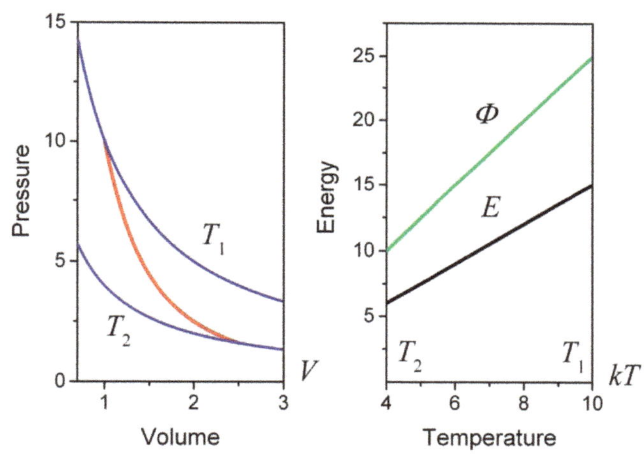

Figure 17.3. Adiabatic process. Left: dependence of enthalpy (green) and energy (black) on temperature. Right: adiabate (red) and isotherms (blue).

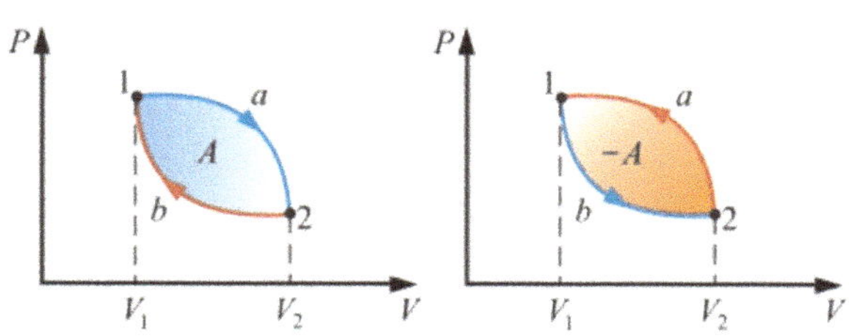

Figure 17.4. Cyclic process in an ideal gas.

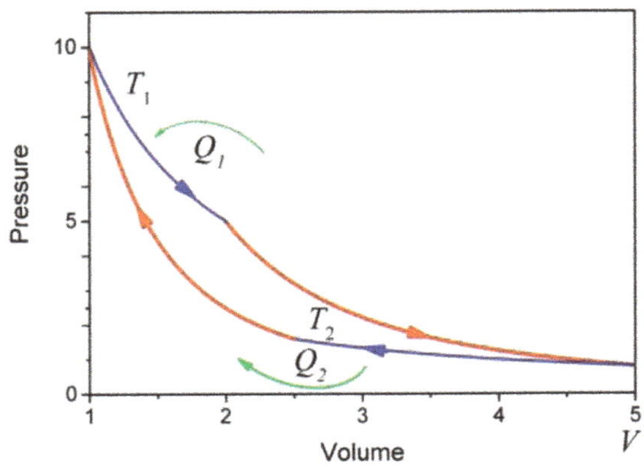

Figure 17.5. Carnot cycle.

of the gas by the amount $\Delta S_1 = Q_1/T_1$ (see table 17.1). Then it is followed by the *isentropic* (reversible adiabatic) *expansion* of the gas (blue). The gas continues to expand, doing work on the surroundings and losing an amount of internal energy equal to this work. The gas expansion causes it to cool to the 'cold' temperature T_2. The entropy remains unchanged. Finally, the reversible isothermal compression of the gas at the 'cold' temperature T_2 (red) makes the surroundings do work on the gas, causing an amount of heat energy Q_2 to leave the system to the low temperature reservoir, and the entropy of the system decreases by the amount $\Delta S_2 = Q_2/T_2$. Finally, once again the isentropic compression of the gas (red) makes the surroundings do work on the gas, increasing its internal energy and causing the temperature to rise to T_1, due solely to the work added to the system. At this point, the gas is in the same state as at the start of step 1 with the same change of the entropy $\Delta S_2 = \Delta S_1$. This implies the fundamental result of Carnot:

$$Q_1/T_1 = Q_2/T_2, \tag{17.10}$$

This means that all reversible heat engines working at a constant temperature produce the same amount of work (the area inside the curvilinear quadrilateral) for a given heat flux from hot to cold. None of them can be more effective since the work in such a process is maximal. Also, entropy measures irreversible changes in a system.

Problem 17.1. The equivalent way of writing the work done in an infinitesimal Carnot cycle follows from the equation $d(dE) = 0$. Prove this from the Maxwell relation $\left(\frac{\partial S}{\partial V}\right)_T = \left(\frac{\partial P}{\partial T}\right)_V$.

Further reading

Adkins C J 1983 *Equilibrium Thermodynamics* (Cambridge: Cambridge University Press)

IOP Publishing

Principles of Statistical Physics and Numerical Modeling

Valeriy A Ryabov

Chapter 18

Processes in a non-ideal gas

We saw in chapter 15 that for high pressure or low temperature the gas molecules become closer and closer together. It makes the interaction between them significant so the kinetic theory of gases as stated above is no longer valid. Of course, all results obtained in the previous chapter may be reproduced in this case simply by the replacement of the equation-of-state of (16.1) with equation (16.2).

Therefore, in this chapter, we shall consider other important examples of stationary *irreversible* processes in which non-ideality is essential. One is the *process of Joule–Thomson*. It involves a slow flow of gas through a porous membrane under a constant difference of pressure without heat exchange (see figure 18.1). Stationarity means here the constancy of pressures.

The initial volume of gas V_1 at a temperature T_1 passes through the barrier, distributed over the volume V_2, obtaining the temperature $T_2 \neq T_1$. The change of energy $E_2 - E_1$ is equal to the difference of works of the compressor to oust particles from volume V_1, equal $P_1 V_1$ and work $P_2 V_2$ on expansion of gas in the right chamber, so,

$$E_1 - E_2 = P_2 V_2 - P_1 V_1. \tag{18.1}$$

Thus, in this process the enthalpy remains constant

$$\Phi_1 = E_1 + P_1 V_1 = \Phi_2 = E_2 + P_2 V_2. \tag{18.2}$$

For an ideal gas, it should be $\Delta T = T_2 - T_1 = 0$ and, consequently, the effect of Joule–Thomson is absent. Therefore, the effect is observable only in a *non-ideal gas* whose atoms, being on average far enough apart, interact weakly with one another, but ‘the friction’ of gas in the membrane leads to a change of temperature.

Consider first the *integral effect of Joule–Thomson*, when the difference of temperatures ΔT is considerable due to the large difference of the pressures $\Delta P = P_1 - P_2$ given $P_2 \ll 1$. Under these conditions the gas, having been non-ideal in the first chamber, expands into the second chamber of large volume $V_2 \gg V_1$

doi:10.1088/978-0-7503-1341-4ch18

becoming an ideal gas since the particles, on arriving there, practically cease to interact. As a result, the temperature in the second chamber is defined by the initial value of enthalpy in the first chamber $\Phi_1 = 2NkT_1$ (recall that 2D calculations are considered) conserving the process of dynamics.

For the numerical evaluation of the effect, we will use the results of chapter 15 where the enthalpy $\Phi_1(T_1, V_1)$ was calculated along isotherm $kT_1 = 0.18$. The results of the calculation are shown in figure 18.2. It is seen that the particles flowing into the second chamber can heat up (a *positive effect* of Joule–Thomson) and be cooled (a *negative effect* of Joule–Thomson). The principle of the work of the refrigerator is just based on the last effect. The change of sign of ΔT occurs at some value $T_1(V_1)$ (dots) which in the limit of large V_1 tends to the *temperature of inversion* T_i (dashed-dotted line).

The obtained results can be explained by means of the van der Waals equation of (16.2). Taking into account equation (16.11), write the condition of equality of an enthalpy for a real 3D gas in the left chamber and an ideal one in the right:

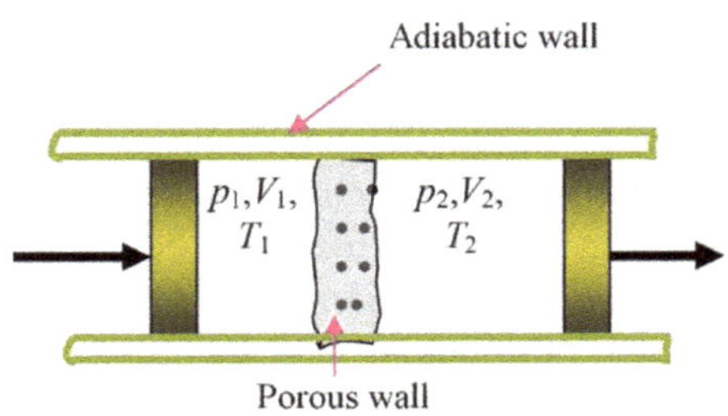

Figure 18.1. The Joule–Thomson effect.

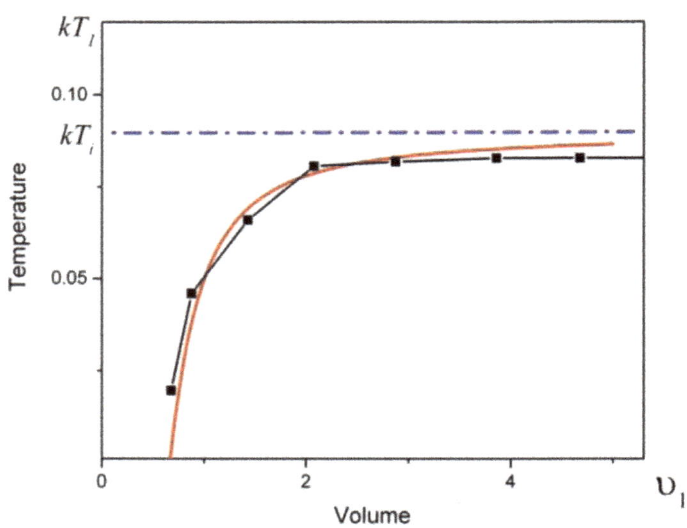

Figure 18.2. Dependence of the temperature T_1 of the volume V_1 for a 2D non-ideal gas. Dots: exact calculation; red curve: van der Waals approximation; dashed-dotted line: temperature of inversion.

$$c_V T_1 - \frac{a}{v_1} + T_1 + T_1\frac{b}{v_1 - b} - \frac{a}{v_1} = c_V T_2 + T_2, \tag{18.3}$$

where $v = V/N$, $c_V = 3/2$ (for a van der Waals 2D gas for $c_V = 1$ this is shown by the red curve in figure 16.1). Then the difference of temperatures in the first and second chambers becomes

$$\Delta T = T_2 - T_1 = \frac{1}{c_P}\left(T_1\frac{b}{v_1 - b} - \frac{2a}{v_1}\right). \tag{18.4}$$

Since $C_P > 0$, the sign of ΔT depends on the sign of the expression in square brackets of this equation, thus causing the change of the sign of the Joule–Thomson effect at a temperature

$$T_i = \frac{2a}{b}\left(1 - \frac{b}{v_1}\right). \tag{18.5}$$

The results of this chapter could be used for the treatment of another *differential effect* of Joule–Thomson. It determines the temperature increment for a small change of the pressure. To proceed further, divide by dP the equation for the enthalpy of equation (14.2) $d\Phi = TdS + VdP$ while holding temperature constant:

$$\left(\frac{\partial \Phi}{\partial P}\right)_T = T\left(\frac{\partial S}{\partial P}\right)_T + V. \tag{18.6}$$

Taking into account the cycle rule (see problem 14.2) the partial derivative on the left transforms to

$$\left(\frac{\partial \Phi}{\partial P}\right)_T = -\left(\frac{\partial \Phi}{\partial T}\right)_P\left(\frac{\partial T}{\partial P}\right)_\Phi = C_P\mu_{JT}. \tag{18.7}$$

Here the *isothermal Joule–Thomson coefficient* $\left(\frac{\partial T}{\partial P}\right)_\Phi = \mu_{JT}$ was introduced. With the help of the *Maxwell relation* (see problem 16.1) the partial derivative on the right of equation (18.6) becomes

$$\left(\frac{\partial S}{\partial P}\right)_T = -\left(\frac{\partial V}{\partial T}\right)_T. \tag{18.8}$$

Now it is expressed through the coefficient of thermal expansion α:

$$\left(\frac{\partial V}{\partial T}\right)_T = \alpha V. \tag{18.9}$$

Replacing these two partial derivatives in (18.7) and (18.8) yields

$$\mu \equiv \left(\frac{\partial T}{\partial P}\right)_\Phi = \frac{V}{C_P}(\alpha T - 1). \tag{18.10}$$

For an ideal gas $\alpha = 1/T$ and this coefficient vanishes. By measuring the Joule–Thomson coefficient, the direction of temperature change can be determined. Not all gases undergo a cooling effect upon expansion. Some gases, such as hydrogen and helium, will experience a warming effect upon expansion under conditions near room temperature and pressure.

In this irreversible process of expansion entropy increases. This process *of entropy increase* is easy to understand assuming that $V_2 \gg V_1$, i.e., the expansion of gas happens essentially into a vacuum. While its energy E remains constant and the total volume varies only slightly, the change of temperature is defined by a derivative $\left(\frac{\partial T}{\partial V}\right)_E$. Hence taking into account equations (8.21) and (8.23) one obtains

$$\left(\frac{\partial S}{\partial V}\right)_E = \frac{\left(\frac{\partial E}{\partial V}\right)}{\left(\frac{\partial S}{\partial E}\right)} = \frac{P}{T} > 0. \tag{18.11}$$

It should be noted that an irreversible increase in entropy is not a property at the microscopic level since motion equations are reversible in time. From this point of view, the change of entropy sets *a direction of time*, prohibiting reverse time in systems.

Problem 18.1. Find the Joule–Thomson coefficient using the van der Waals equation.

Further reading

Zemansky M W 1968 *Heat and Thermodynamics; An Intermediate Textbook* (New York: McGraw-Hill)

IOP Publishing

Valeriy A Ryabov

Chapter 19

Reaction coordinate

In processes of chemical dissociation, the diffusion of defects in a solid and in structural phase transitions, the behavior of one selected degree of freedom of the ensemble may be a matter of great importance. Under the influence of thermal motion, a coordinate correspondent to this degree of freedom can displace spontaneously from one equilibrium position to another, crossing a potential barrier region. One of the representative examples of such a behavior is the jump of an atom A to a neighboring vacant lattice site B (the vacancy) as is shown in figure 19.1. The animation in figure 19.2 illustrates this phenomenon in dynamics. The trajectory of this jump, the *reaction path* (shown by scatters) consists of a continuous sequence of *coordinates of reaction.* Apparently, other trajectories in the vicinity of the reaction trajectory are also possible under the influence of thermal motion. Determine the probability of this jump.

As is known from mechanics, the equilibrium positions of atoms at zero temperature (without thermal vibrations) are defined by the minimum of interaction potential. To drag an atom from one equilibrium position to another in a lattice, it is necessary to overcome a potential barrier. All points around the path AB form a potential profile of the *saddle-point configuration* (figure 19.3). Note that during a jump the atoms surrounding the moving atom are displaced from lattice cells. Finding these displacements corresponding to a total potential minimum for each point of a trajectory is not trivial even for $T = 0$. Details of a special procedure of a relaxation of a lattice are given in appendix A.12. Calculation of the potential profile in a hexagonal lattice with Lennard-Jones potential (0.1, 1.0) and the distance $d = 1$ between the nearest neighbors, gives the height of the potential barrier $U_b \approx 0.37$.

At finite temperature, at first sight, it might seem that this problem should be solved similarly to the case $T = 0$ above: by finding an average profile of the total energy E along the reaction coordinate for a given temperature. However, it is not so. It is evident from the definition of the energy of equation (11.12) that a change in any microscopic average parameter of the system should be accompanied by

doi:10.1088/978-0-7503-1341-4ch19

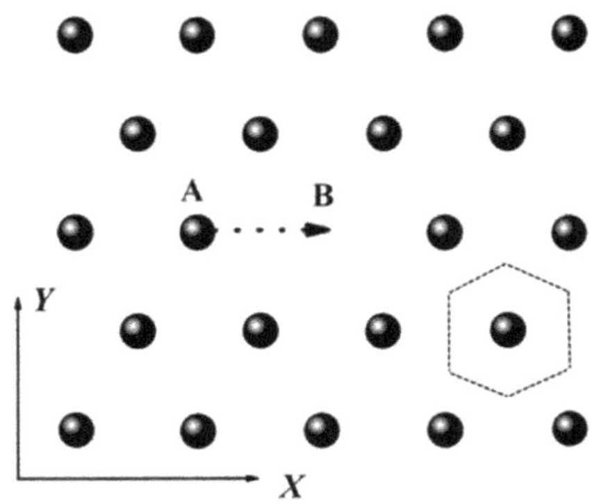

Figure 19.1. The reaction path is the jump of atom A into an unoccupied lattice site, position B (vacancy).

Figure 19.2. Vacancy jump in vibrating lattice. Animation available at http://iopscience.iop.org/book/978-0-7503-1341-4.

changes of the phase volume of the system and, hence, the entropy. Therefore, in general, only the profile of free energy determines the equilibrium state of the system, while the reaction coordinate displaces. However, looking ahead we will see that the main feature of a jump, namely the temperature dependence of its probability, is well reproduced by a static computation.

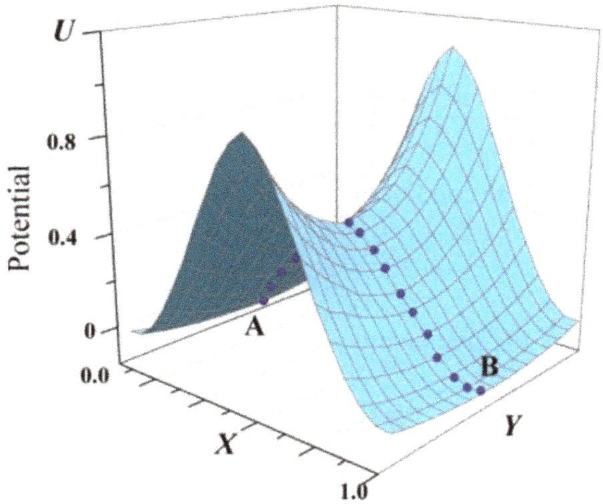

Figure 19.3. Potential profile near the path of the jump (shown by blue circles).

According to equations (6.1) and (9.10), the probability of a particle coordinate x_i having the mean value q is equal to

$$P(q) = \langle \delta(x_i - q) \rangle = \frac{1}{Q} \int d\mathbf{x} d\mathbf{p} \delta(x_i - q) \, e^{-\beta \mathcal{H}(\mathbf{x}, \mathbf{p})}. \tag{19.1}$$

Hence the *profile of free energy* is defined as

$$F(q) = -\frac{1}{\beta} \ln P(q). \tag{19.2}$$

Prove that this equation reproduces the potential profile well in static computations ($T = 0$). For this purpose consider the increment of free energy caused by the change of the reaction coordinate $q_1 \to q_2$

$$F(q_2) - F(q_1) = \int_{q_1}^{q_2} dq \frac{\partial F(q)}{\partial q} = -kT \int_{q_1}^{q_2} dq \frac{1}{P(q)} \frac{\partial P(q)}{\partial q}. \tag{19.3}$$

Further write the integrand here in the form

$$\frac{1}{P(q)} \frac{\partial P(q)}{\partial q} = \frac{\frac{1}{Q} \int d\mathbf{x} d\mathbf{p} \, e^{-\beta \mathcal{H}} \frac{\partial}{\partial x_i} \delta(x_i - q)}{\langle \delta(x_i - q) \rangle}. \tag{19.4}$$

Carrying out the integration in parts in the numerator, we obtain

$$\int d\mathbf{x} d\mathbf{p} \, \frac{\partial e^{-\beta \mathcal{H}}}{\partial x_i} \delta(x_i - q) = -\beta \int d\mathbf{x} d\mathbf{p} \, e^{-\beta \mathcal{H}} \frac{\partial \mathcal{H}}{\partial x_i} \delta(x_i - q). \tag{19.5}$$

The quantity

$$\frac{1}{P(q)}\frac{\partial P(q)}{\partial q} = -\beta\frac{\left\langle \frac{\partial \mathcal{H}}{\partial q}\delta(x_i - q)\right\rangle}{\langle \delta(x_i - q)\rangle} \tag{19.6}$$

looks like an average of the conditional probability of coordinate x_i being equal to q

$$\langle \dots \rangle_q^c = \frac{\langle (\dots)\delta(x_i - q)\rangle}{\langle \delta(x_i - q)\rangle}. \tag{19.7}$$

As a result the gain of the free energy

$$F(q_2) - F(q_1) = \int_{q_1}^{q_2} dq \left\langle \frac{\partial \mathcal{H}}{\partial x_i}\right\rangle_{x_i=q}^{c} = \int_{q_1}^{q_2} dq \left\langle \frac{\partial U}{\partial x_i}\right\rangle_{x_i=q}^{c} \tag{19.8}$$

is determined by the average force acting on a particle, being in a fixed position $x_i = q$. Integrating with respect to a variable q, we find that the difference of free energies coincides with a difference between mean values of potential energy in points q_1 and q_2. The jump through the barrier occurs when the coordinate of reaction reaches the saddle-point corresponding to the barrier height $U(q_m) = U_b$, so the difference between the free energies of the ground and barrier states of equation (19.8) becomes equal to the *activation* (*migration*) *energy* of the process of the migration of an atom

$$\varepsilon_a = \langle U(q_m)\rangle - \langle U(q_A)\rangle. \tag{19.9}$$

Since the displacements of the equilibrium positions due to thermal expansion (see chapter 7) are usually small, a good approximation for the average of equation (19.9) will be the value of the potential of a lattice without thermal fluctuations, but taking into account elastic strain near a defect (*relaxation energy*). As a result, the difference of free energies in equation (19.9) can be defined by minimization of the lattice potential: $\partial U/\partial x_i = 0$, $i = 1, N$ in both positions (see appendix A.7).

Eventually the *probability of jump* from the initial equilibrium position q_0 to the saddle-point q_m, according to equation (19.4), can be written in the form

$$\frac{P(q_m)}{P(q_0)} = \exp(-\beta\,\varepsilon_a). \tag{19.10}$$

Apparently, the densities of particles at the top of a barrier and in the bottom of the potential well are in the same ratio. Keeping in mind that the probability of a jump looks like $P \sim 1/\tau$, where τ is the average time of a jump, then, taking into account equation (19.10), this time can be presented as

$$\tau = \tau_0 \exp(\beta\,\varepsilon_a), \tag{19.11}$$

where the pre-exponential factor τ_0 is in the order of the period of thermal vibrations of atoms in the lattice.

The probability of a jump can be determined numerically from MD simulation (part of the MD run is shown in the animation). For this purpose, we shall use the

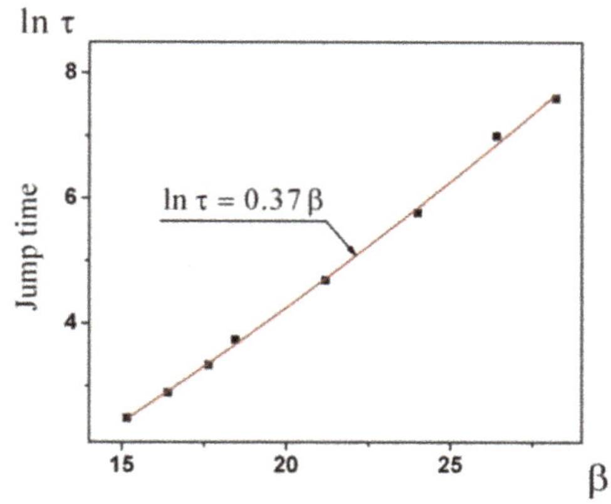

Figure 19.4. The dependence of the duration of the jump along the reaction coordinate on inverse temperature. The red line is the result of interpolation by the exponential distribution.

NVE ensemble described previously in chapter 5. The system of $N = 799$ atoms vibrating in the 2D hexagonal lattice and interacting via the same Lennard-Jones potential is considered. The current time of the jump was fixed when one of six atoms surrounding the vacancy passed to the vacant cell and thermalized there. In terms of the calculation algorithm, this process occurs in the moment of the output of an atom from the area per atom. This area (the *Wigner–Seitz cell*) is the hexagon marked by the dashed lines in figure 19.1.

The temperature dependence of the average duration of a jump of the atom (see appendix A.7) is given in figure 19.4. It is seen that the slope of the line in the logarithmic scale with respect to reciprocal temperature $\beta = 1/kT$ coincides with the value of the activation energy calculated from a potential profile along the reaction path in figure 19.3: $U_b \approx 0.37$. This curve (*Arrhenius's law*) is typical for all processes whose probability depends on the thermal vibration of atoms.

We will see in the following chapter that the method of *thermodynamic integration* derived above may be applicable not only to a trajectory of the reaction of a separate particle in the ensemble, but also to the entire ensemble as a whole (compare also with equation (11.16)). Consider this in more detail. Let the potentials of the two ensembles at given temperature be equal to U_0 and U_1. Then from equation (19.8) a small increment of potential $U_1 = U_0 + \delta U$ yields

$$\delta F = \langle \delta U \rangle_0. \tag{19.12}$$

Using this ratio it is possible to carry out the thermodynamic integration between two unnecessary close values of potentials, considering a series of intermediate states—perturbations depending on any parameter of an order a: $U_\lambda = [1 - a(\lambda)]$ $U_0 + U_1$ with $a(\lambda) = 0 \div 1$, so

$$F_1(\beta) - F_0(\beta) = \int_0^1 d\lambda \left\langle (U_1 - U_2)\frac{\partial a}{\partial \lambda} \right\rangle_\lambda . \tag{19.13}$$

The parameter of an order $a(\lambda)$, in principle, can depend on all coordinates of the system.

In conclusion, one essential remark should be made. We have seen how the mobility of a defect can be described in terms of a thermodynamic treatment. However, the mobility or activation energy itself is unable to predict the equilibrium characteristics of defects as an ensemble. For example, an equilibrium concentration of defects or their clusters is determined by another characteristic, a *formation energy*. For the vacancy, this value needs to be equal to the energy necessary for the atom to escape from the lattice site to a surface. The calculation of this value is very similar to what we have done here for the migration energy.

Further reading

Tuckerman M E 2010 *Statistical Mechanics: Theory and Molecular Simulation* (New York: Oxford University Press)

IOP Publishing

Principles of Statistical Physics and Numerical Modeling

Valeriy A Ryabov

Chapter 20

Structural phase transitions

The qualitative change of the macroscopic properties of a homogeneous system due to a changing control parameter is called a *phase transition*. Sometimes, a phase transition is related to a change in the degree of order in the system quantified by the *order parameter*—spontaneous symmetry breaking. A typical example of such behavior is a spontaneous change of lattice symmetry owing to *structural phase transitions*. In this case, the order parameter is a deformation. Consequently, the reaction coordinate which undergoes a spontaneous change relates not to a separate particle, but to the entire ensemble as a whole. As in the case of a defect jump, the system has to cross a barrier in order to transform from one structure to another. In practice, this barrier is rather large to observe the transition within the accessible simulation time. Therefore, we will use below a system close to the point of mechanical instability by an appropriate choice of interatomic potential.

In particular, we consider the transition from a square to the hexagonal lattice. It occurs by means of an extension along one of the diagonals of a rhombus and expansion along the other (figure 20.1). The atomic density upon this transition changes from $\rho = 1$ to $\rho = 2/\sqrt{3}$. Therefore, it is natural to choose parameters of uniaxial deformations, stretching in the X- and Y-directions, as the reaction coordinates for this transition. Since we are dealing with the change of symmetry

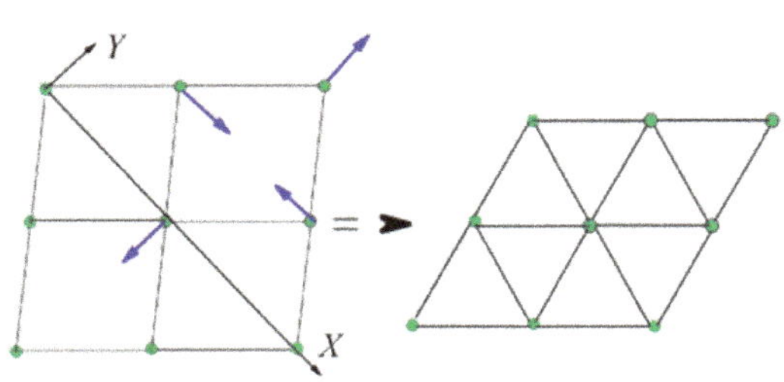

Figure 20.1. Deformation of a square lattice into a triangular lattice. The deformation directions of the square lattice already being slightly deformed are indicated by the blue arrows on the left.

doi:10.1088/978-0-7503-1341-4ch20

of the system, the deformation here acts as an *order parameter*. It is seen from figure 20.1 that the use of a pair potential in the nearest-neighbor approximation makes the square lattice unstable with respect to a pure shear. Indeed, it does not change the total energy of a lattice, since the distances between the nearest neighbors in the rhombus do not change. The evolution of such system will inevitably lead to the formation of a hexagonal lattice, because an interaction between atoms along a small diagonal of a rhombus 'turns on' as soon as its length becomes closer to the distance of the minimum of pair potential. To provide some stability of the square lattice at low temperatures, we introduce a slight repulsion at the expense of a next-to-nearest-neighbor interaction at the distance $r_2 = \sqrt{2}$ in a rectangular (square) lattice (see figure 20.2). The small barrier in point r_m keeps the atoms of the second coordination circle located at the distances ~ r_2 from collapse into a hexagonal lattice.

Figure 20.3 shows the potential profile of lattice potential $U(\lambda_x, \lambda_y) = U(\{\mathbf{r}'_i\})$ as a function of two uniaxial deformations of the reference square lattice with atom coordinates $\mathbf{r}_i$ and $\lambda_x = \lambda_y = 1$ to the current deformed lattice $\mathbf{r}'_i$: $x'_i = \lambda_x x_i$ and $y'_i = \lambda_x y_i$. The reaction trajectory (this dotted line indicates the most probable path) gives a representation of a process of lattice structure transformation during phase transition. Starting from point A the square lattice with $(\Delta\lambda_x, \Delta\lambda_y) = (0, 0)$ has to overcome the barrier for transforming to the hexagonal lattice in the point B with $(\Delta\lambda_x, \Delta\lambda_y) = (-0.293, 0.225)$.

Similar to the activation jump considered in the previous chapter, the lattice during transition goes from one equilibrium position in point A with the potential energy $U^{\text{sq}}_{\text{min}} = -0.331$ to another with $U^{\text{hex}}_{\text{min}} = -0.530$ in point B (the energies are measured from barrier energy).

Using the argument of the previous chapter, it is clear that if an initial temperature of the square lattice is much less than the height of the barrier

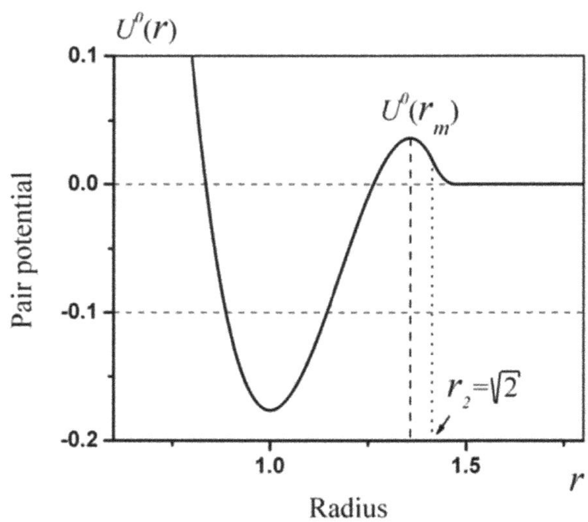

Figure 20.2. Pair interaction potential for the transition from a square lattice to a triangular lattice.

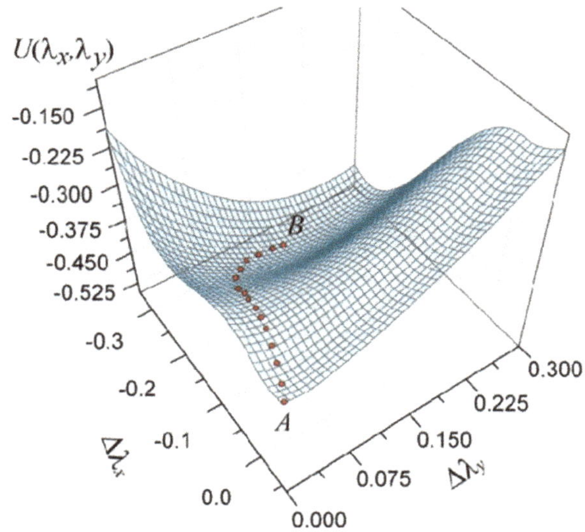

Figure 20.3. Potential landscape near the reaction path for the phase transition (scatters). A: a square lattice, B: a triangular lattice

$kT_{\rm sq} \ll \varepsilon_a = |U^{\rm sq}_{\rm min}|$, the square lattice is stable. With temperature increase, the probability of transition $\sim \exp(-\frac{\varepsilon_a}{kT})$ becomes sufficient to overcome the barrier in process of transition.

To understand how this temperature induced transition evolves in time, carry out an MD simulation on a 2D system. As we have seen in chapter 7, the uniaxial deformation can be included in the dynamics as an independent variable. In this case, the joint dynamics of the particles and stretching λ_x and λ_y in two mutually perpendicular directions X and Y obeys equation (1.25)[1]. The volume changes as $V = V_0\lambda_x\lambda_y$.

Calculations in terms of the isobaric–isoenthalpic ensemble described in chapter 13 were carried out with the help of the equations of motion of (1.25). $N = 900$ atoms in an initial 2D square lattice interact via the pair potential shown in figure 20.2. Starting from a small kinetic energy the temperature is gradually increased with the increment $\Delta kT/\tau = 2 \cdot 10^{-7}$, as was done in chapter 17. Details of the calculation procedure are presented in appendix A.13. While the kinetic energy/temperature is increasing to a value comparable with the activation energy (the barrier's height in figure 20.3) the initial phase becomes unstable, triggering a transition. Figure 20.4 shows the time evolution of instantaneous temperature and stretching along the axes X and Y.

The abrupt increase of the kinetic energy at the moment of transition $t \sim 310$ is explained by the release of a certain amount of heat, which is called the *heat of phase transition.* Its origin seems evident if you take into account the energy release

[1] Strictly speaking, except for stretching, possible 2D deformations also include shear. However, for the considered transition the reaction coordinate is nearly along a single direction and the correspondent correction is small.

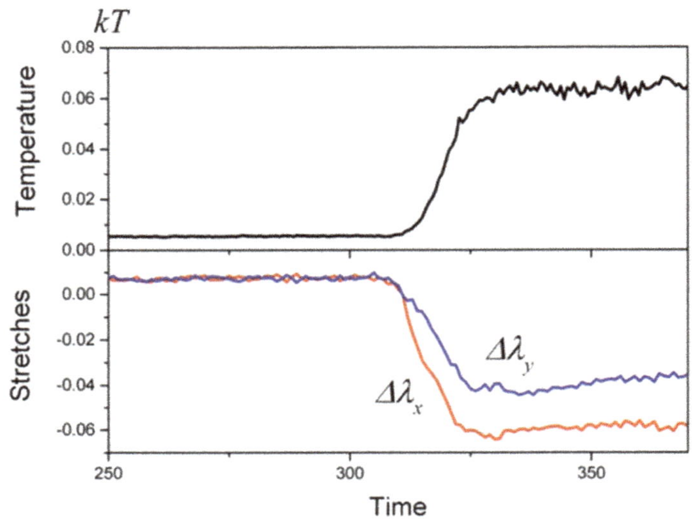

Figure 20.4. Time evolution of the lattice parameters during phase transition. Top: instantaneous temperature; bottom: uniaxial stretches along the axes *X*, *Y* and volume (black).

$U_{\min}^{\mathrm{hex}} - U_{\min}^{\mathrm{sq}} < 0$ equal to the difference between the potential energies after and before transition.

The change of all deformation parameters indicates a restructuring of the lattice. Here we make a small digression. Although it might seem natural to try to characterize the atomic arrangement in terms of a set of distances between the atoms, the spatial distribution functions in a system contain a considerable amount of information about the local structure and the fluctuations. The atoms are constantly in motion, their coordinates are fluctuating. Hence, a more appropriate measure of the atomic structure is a distribution function of distances of atoms $\mathbf{r}_{ij}$ from each other, the *pair correlation function.* We already discussed a distribution like this in chapter 3. Built on this function, the more simple *radial distribution function* $g(r)$ gives the localization of these distances and the number of neighbors in different coordination spheres, regardless of the mutual angular orientation (see appendix A.13).

Figure 20.5 demonstrates the structure in terms of the radial distribution function before (dashed curve) and after transition (solid curve). In the square lattice, the peaks correspond to the location of four nearest $r = 1$ and four next-to-nearest neighbors $r = \sqrt{2}$. After transition, the number of nearest neighbors in the point $r = 1$ increases up to six. But the position of the six next-to-nearest neighbors now appears in the point $r = \sqrt{3}$. The variance of the distributions is the consequence of the thermal displacements of atoms and, as we will see, of defects of the crystal lattice.

Although the radial distribution function reveals the expected structural changes resulting from transition, it does not explain a discrepancy in a magnitude of deformation. It is seen from figure 20.4 that the final uniaxial deformation parameters are rather small compared with the values following from the pure potential consideration above. The reason for this can be understood if we look at the time evolution of a specific arrangement of atoms during transition (see figure 20.6).

The transition occurs not in the whole volume but simultaneously in several blocks, *droplets*, oriented in an arbitrary manner with respect to the initial lattice (see figure 20.7). Homogeneous nucleation of areas of new phases of larger atomic density causes *discontinuities, emptinesss* and extensive linear defects such as *stacking faults*. Because of these structural defects, the change of density turns out to be much

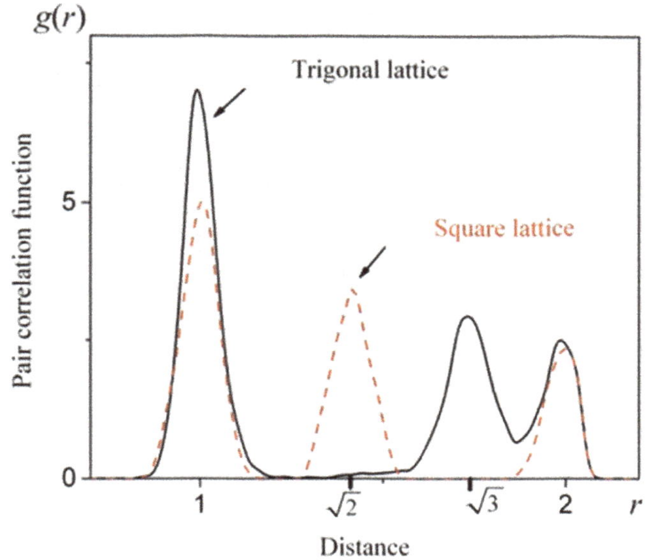

Figure 20.5. Change of the radial distribution function after phase transition.

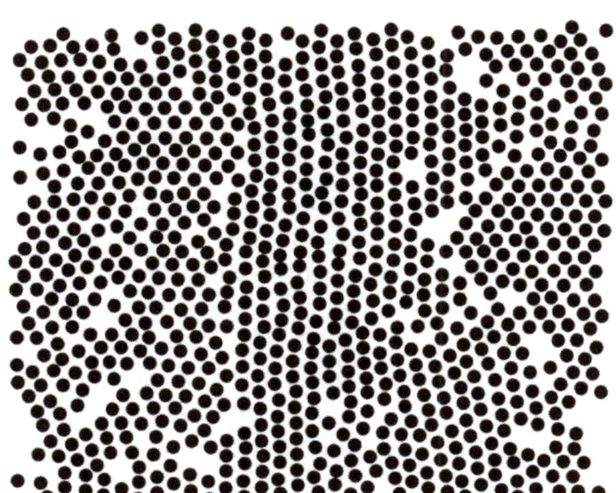

Figure 20.6. Lattice after structural phase transition.

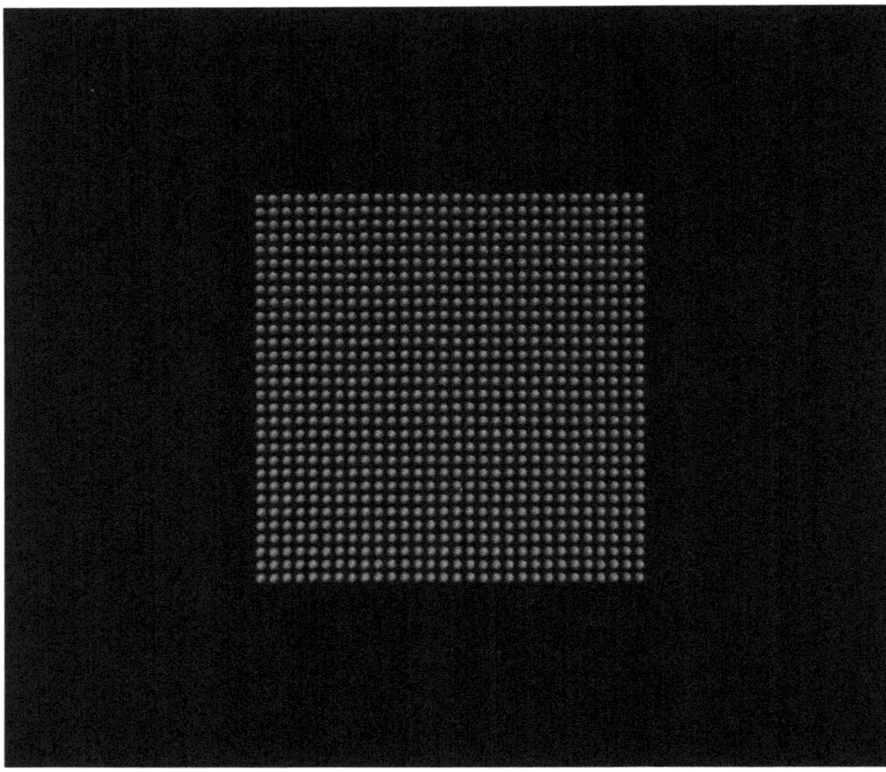

Figure 20.7. Structural phase transition. Animation available at http://iopscience.iop.org/book/978-0-7503-1341-4.

less than the value $\rho \sim 2/\sqrt{3}$, corresponding to a perfect hexagonal lattice. Apparently, the minimum of free energy relating to the final structure and macroscopic parameters does not match the pure potential treatment. Observed changes correspond to *first order phase transition*, as is usually inherent to the structural phase transition.

The transition point is characterized by equality of the thermodynamic potentials of both phases, as it must be for equilibrium. In addition, the process of overcoming the potential barrier is accompanied by a change of point symmetry of a lattice in the course of transition: from the spatial rotation group C_4, inherent in a square lattice, to group C_6 for a hexagonal lattice. This causes a change of the *configuration entropy* $\Delta S = kN \ln(n_2/n_1)$, where $n_1 = 1$ and $n_2 = 2$ are the numbers of atoms in the unit cells of the first and second phase, respectively. This contribution usually prevails over the vibrational (thermal) one. The abrupt change of volume and entropy, according to equation (14.4), leads to a discontinuity in the derivatives of of the Gibbs potential on temperature and volume $(\partial G/\partial T)_P = -S(\partial G/\partial P)_T = -V$, leaving continuous their values at the transition point.

In each droplet, the transition occurs along the potential path, but because of the lack of any adjustment of the internal pressure to external, the thermodynamically favorable path is an independent nucleation of droplets in a volume of the matrix depending on their surface energy. Typically, during the phase transition, the system is passing through a series of intermediate states with insignificant minima of free energy, to a global minimum corresponding to a stable structure. In practice, this structure may be revealed by special MD methods based on artificial shaking of deformation parameters to overcome these insignificant barriers and suppress heterogeneous nucleation.

The considered transition is only one of many examples of a fairly wide variety of structural first order phase changes. The mentioned features of the ensemble evolution are inherent also in the 3D case, which also shows the fluctuation model of nucleation of a new phase through a stage of formation of droplets. All mentioned peculiarities make the theoretical (but not numerical) analysis of phase transitions rather difficult in terms of free energy with respect to a parameter of a certain order. However, in statistical physics, there is a rather simple method for studying phase transitions, which is not connected with classical mechanics at all (see chapter 22).

Further reading

Marton R, Laio A and Parrinello M 2003 Predicting crystal structures: the Parrinello–Rahman method revisited *Phys. Rev. Lett.* **90** 075503

IOP Publishing

Principles of Statistical Physics and Numerical Modeling

Valeriy A Ryabov

Chapter 21

Phase transitions with change of aggregate state

As we have seen in chapter 16, a change of temperature and pressure cannot be reduced only to a change of the atomic density (see figure 16.1). It may entail the change of an *aggregate state* of substance: a solid, liquid, gas. Let us discuss these changes in more detail. In the thermodynamic process of the *first order phase transition,* the density, compressibility, specific heat and other system parameters change abruptly. From the point of view of dynamics this transition occurs with absorption (release) of a certain amount of *heat of phase transition*. In order for the phase transition to continue, it requires taking away (or bringing) this heat continuously from the body. Hence the microcanonical ensemble where the temperature is established as a result of dynamics and, therefore is not quite predictable, is not good enough to trace the changes in thermodynamic potentials. In specific calculations, it is more convenient to use the *isothermal NVT ensemble* discussed in the previous chapters or even the *NPT* ensemble in which additionally a constant pressure is maintained.

At first, we shall use the results of the isothermal ensemble obtained earlier in chapter 16 for a non-ideal gas. Figure 21.1 shows schematically the instant arrangement of atoms in the 2D ensemble at three various temperatures corresponding to three points in the equation-of-state of (16.2) shown in figure 16.1: $kT_1 = 0.11$, $kT_2 = 0.12$, $kT_3 = 0.17$ for the same pressure $P = 0.12$. In the first state (the left panel), at a low temperature, the system is represented by a hexagonal lattice. Then, with temperature increase, the ordered arrangement of atoms of a lattice is replaced by a disordered *liquid state* with a minor change of density (middle panel) and extremely low compressibility. Finally, at sufficiently high temperature, atoms become almost free, despite the attraction between them, forming *vapor*.

The insets in figure 21.1 show the behavior of the radial distribution function for all three phases. In a hexagonal lattice, the number of neighbors at distances $r_1 = 1$ $r_2 = \sqrt{3}$ and $r_3 = 2$ is equal to six. They give three well-separated peaks in

doi:10.1088/978-0-7503-1341-4ch21

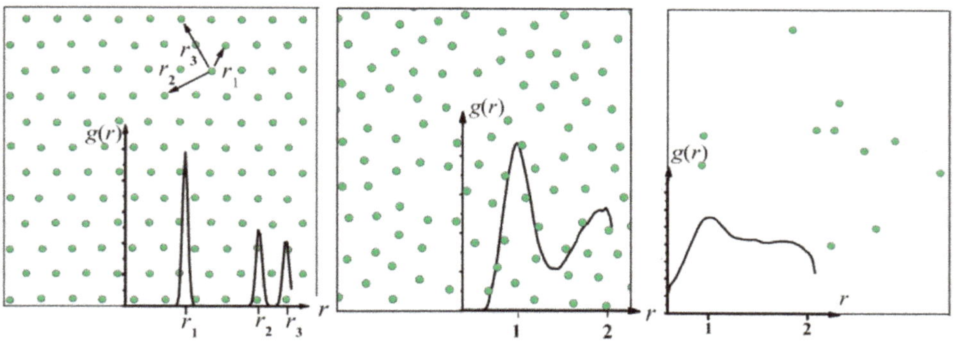

Figure 21.1. The temperature phase transitions in a 2D ensemble. With increasing temperature: left, a triangular lattice; middle, the liquid state; right, non-ideal gas. All panels refer to the same volume of 10 × 10. The insets show the radial distribution function.

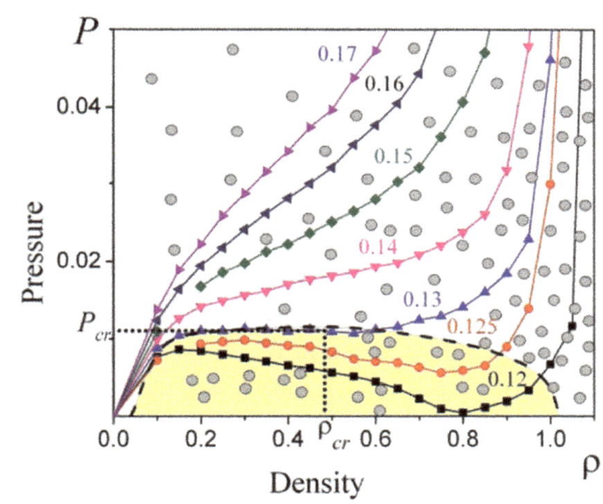

Figure 21.2. Areas of phase separation

function $g(r)$. In the amorphous, liquid state, there is a well-isolated peak near $r = 1$ followed by a minimum. Those are associated with an attraction minimum in pair potential and repulsion at a strong rapprochement of atoms (*short-range order*). However, the regularity of the atomic arrangement is no longer present and the fluctuating density is close to an average. Finally, in the vapor state, the mutual arrangement of atoms, being close to completely disordered, is characterized by uniform density except for a decrease in a region of strong repulsion. All listed states may coexist in various proportions depending on the macroscopic parameters.

More detailed information about coexisting thermodynamic phases is given by the equation-of-state shown in figure 21.2. This figure differs from figure 16.1 by scaling over the pressure. The isotherms with a temperature lower than some *critical temperature* T_c form a region with *negative compressibility*, where the isotherm slope is $\partial P/\partial V < 0$ (marked in yellow). Note that the van der Waals equation of (16.2) represents the algebraic equation of the third order with respect to ρ having three roots (in the figure, this area is marked in yellow). In the *critical point* (P_c, ρ_c) (dashed line) the magnitudes of all three roots coincide.

Because of this, the value of density ρ has a gap typical for a first order phase transition. The gap area (marked by points on isotherms) corresponds to a mixed state *vapor–liquid*. Its position on the left of the *critical density* ρ_c was determined in calculations visually: by the disappearance of *clusters* of a liquid phase which are easily distinguishable by the average distance between particles in them (equal to unity and corresponding to the minimum of pair potential) on the background of particles in a gas phase with a lower density. The locations of points on the right were determined by the merging of clusters into one liquid phase.

To the left of the *critical isotherm* at $T_c = 0.13$, there is a phase of non-ideal gas—a vapor in which, unlike in an ideal gas, there are potential interaction effects. Finally, for small density and high temperature, isotherms, according to the ideal gas law, are close to a straight line. Lower than the temperature $T = 0.13$, there is an area of negative pressure due to the predominance of the virial in the expression of equation (7.8) over its kinetic part. It starts to influence the attraction between particles which leads to gaps of continuity—space between fragments of a condensed phase. An example of such behavior is shown in figure 21.3. This two-phase area marked in figure 21.2 by the dashed line was determined visually by the presence of clusters. At $T = 0.13$, $V = 2.2$ the clusters of a liquid phase are separated by spaces of almost the same size.

To the right of the *critical isotherm* with ascending density, there are areas of liquid, *melting* and crystal phases. More conveniently, all areas of existence and co-existence of various phases can be shown in the *phase diagram* (figure 21.4) collected from calculations shown in figure 21.2. It shows the temperature dependence of the density on the lines separating phases. The difference in arrangement of atoms was tracked by means of the radial distribution function.

Figure 21.4 shows the simulation results for the same *NVT* ensemble and the density $\rho = 0.9$ This *isochore* (shown by empty circles) at *a melting point* $kT_m = 0.034$

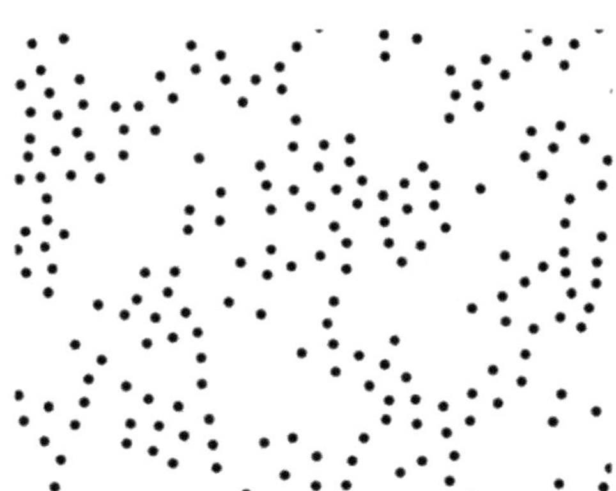

Figure 21.3. Structure belonging to the phase of co-existence of liquid and vapor.

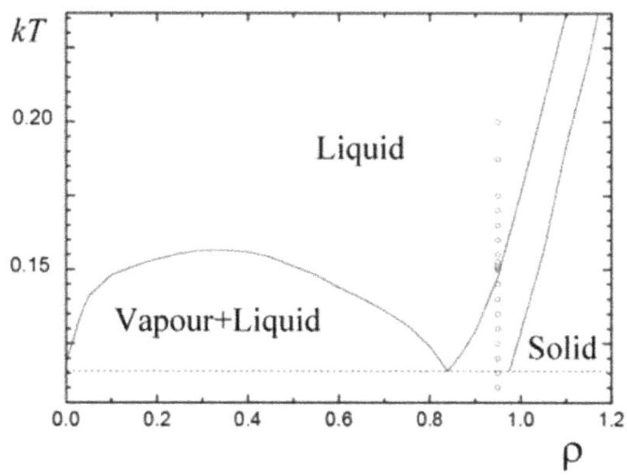

Figure 21.4. Phase diagram of the 2D ensemble.

crosses the line separating the liquid and solid phase. At the melting point, compressibility (equal, for ideal gas, to unity: $\kappa = PV/NkT = 1$) is discontinuous.

The same sort of gap undergoes heat capacity at constant pressure. We already faced a similar discontinuity of specific heat in magnetic systems in chapter 11. A more detailed analysis will be given below in chapter 22.

As a second example, consider the simulation of a *solid–liquid transition* in a realistic substance, lead. For this purpose, we shall use an isothermal–isobaric *NPT* ensemble which is a combination of the two $NP\Phi$ and NVT ensembles considered earlier. The generalization of the calculation algorithm to the case of the isothermal–isobaric ensemble, the most important in practical applications, assumes an extension of the motion equations for particles to include in dynamics the barostat degrees of freedom, taking under control the size and a form of a cell of equation (1.25), and thermostatting of equation (10.4). Such a procedure, in principle, can be formulated without difficulty on the basis of the integrators described in appendices A.7 and A.8. However, practical use of these algorithms gives a low accuracy. More efficient is an operator method of a solution of motion equations by means of a *Liouville operator*. The correspondent *RESPA algorithm* (reversible reference system propagator algorithm) in a simple realization is described in appendix A.14.

A comparison of the initial and final values of the enthalpy of the equilibrium phase formed as a result of the evolution of an isothermal ensemble shows that the equilibrium phase corresponds to the minimum of free energy. Since $F(T, P) = \mu N$, the phase transition occurs when the chemical potentials of the two phases are equated $\mu_{\text{gas}} = \mu_{\text{liq}}$, on a vapor–liquid boundary an exchange of energy, volume and particles can take place. The exchange of energy continues until the temperatures are different (see chapter 6), the exchange of volumes of phases occurs while the pressures are different and particles move from the liquid phase to the gaseous phase while there is a distinction in chemical potentials.

At the critical point $T = T_c$, the gap in the volume, entropy and specific heat C_p disappears, and such an isotherm corresponds to a *second order phase transition*.

Calculation of the melting point in a real substance is not a trivial task. In contrast to the phase transition considered in chapter 20, heat is not only required to

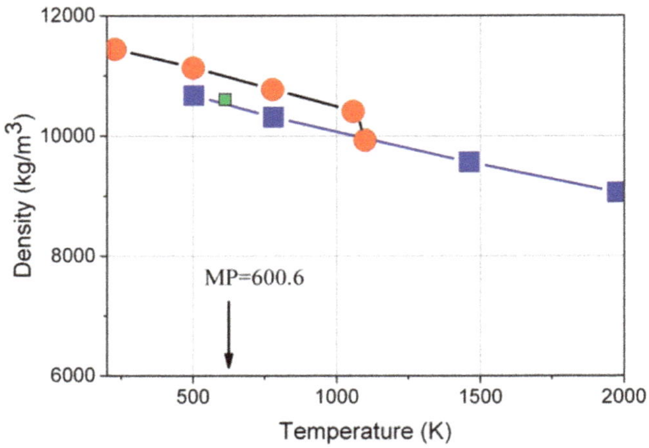

Figure 21.5. Equation-of-state of lead. The experimental melting point (green square) is indicated by the arrow. Red circles: crystal melting; blue squares: liquid melting.

raise the temperature to the melting point. Further heat needs to be supplied for melting to take place: this is called the *heat of fusion* and is an example of *latent heat*. In addition, the solid–liquid transition represents only a small change in volume and, in most cases, a substance is denser in the solid than in the liquid state, so the melting point will increase with increases in pressure. Otherwise reverse behavior occurs. The next important circumstance: the freezing and melting points do not coincide because of their different trajectories in free energy evolution.

Some of these features are seen in figure 21.5. The calculation of equations-of-state was performed for a system of $N = 4000$ atoms interacting via the EAM potential: Pb.lammps.EAM. Details of calculations for $P^{\text{ext}} = 0$ within the NPT ensemble are given in appendix A.14.

The experimental melting point is marked by the arrow. The upper curve is obtained if we start from a crystal lattice, gradually increasing temperature. The lower curve is obtained if we start from a high temperature where just a liquid state takes place.

Problem 21.1. Prove that to maintain equilibrium between the two phases, the small change of pressure and temperature fulfils the $(\frac{dP}{dT})_{\text{eq}} = \frac{\Delta\Phi}{T\Delta V}$ equation (the Clapeyron equation).

Further reading

Sethna J P 2005 Statistical mechanics: entropy *Order Parameters and Complexity* (New York: Cornell University Press)

Chapter 22

Phase transitions in the Ising model

Due to its conceptual simplicity and wide applicability, the Ising model plays a fundamental role in the theory of critical phenomena. In chapter 11, we already met with the calculation of the specific heat of a ferromagnetic, which revealed a singularity at some *critical* temperature T_c. Another characteristic of magnetics, the total magnetic moment or *magnetization*

$$m = \frac{1}{N}\sum_{i=1}^{N} s_i = \frac{N_+ - N_-}{N_+ + N_-}, \tag{22.1}$$

is even more representative from the point of view of details of phase transition. Here $N_\pm$ is the number of spins directed up (down) along one axis Z (assuming a 2D lattice of spins in the plane XY). As for the gas–liquid phase transition, where the density ρ was used to distinguish one phase from another, this variable can distinguish a disordered phase from an ordered phase. In a perfectly ordered state $m = \pm 1$, depending on whether the spins are aligned along the positive or negative Z-direction. Correspondingly, the average total magnetization $\langle m \rangle$ becomes extensively thermodynamically observable.

For numerical simulations consider a 2D square-lattice Ising model in ferromagnetics ($J > 0$). Repeating the calculation procedure described in chapter 11 for this quantity in a lattice of spin 32×32, it is possible to obtain the temperature dependence of magnetization. A completely ordered state at $T = 0$ with all spins being in the same direction, gradually becomes disordered with the growth of temperature (inset in figure 22.1). At $T = T_c = 2.23$, the average spin vanishes with a singularity of the derivative. Thus, the disappearance of the magnetization occurs at the same point as a singularity in the specific heat. At this temperature a first order phase transition takes place, in which the *ferromagnetic state* with $\langle m \rangle \neq 0$ passes into a *paramagnetic* state with $\langle m \rangle = 0$. Thus, the order parameter's role in this transition is magnetization, separating at the point T_c the ordered and disordered

doi:10.1088/978-0-7503-1341-4ch22

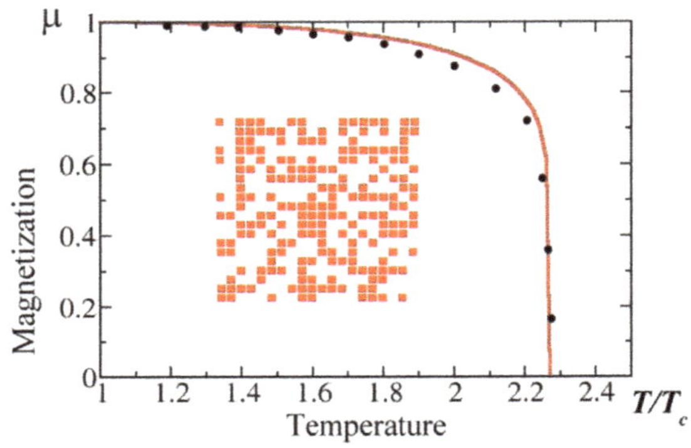

Figure 22.1. Temperature dependence of the magnetization. In the center is shown a piece of the spin lattice at $T = 2.0$, in which one of the directions is marked in red. Points: the mean-field approximation.

phases. For this reason, this phase transition is also an example of *order–disorder transition*.

The clusters of a new phase seen in the inset in figure 22.1 (at $T = 0$ the entire tab would be red) are becoming less and less with growth of temperature. Vice versa, the picture of quenching is very similar to the observed coexistence of the phase shown in figure 21.3. The average size of these clusters is called *correlation length*. It increases with temperature and diverges at the critical point. It means that all sizes of clusters are presented in the system, but there is no net magnetization. Above the transition temperature, the system has no global order, but it still contains ordered clusters. In a limit of high temperatures, the correlation length tends to zero, and any order disappears.

As well as in the previous examples, the first order phase transition is accompanied by a jump of derivatives of m and heat capacity in *Curie point* T_c. The fit of the curve in figure 22.2 as a critical point is given by a function

$$m \sim (T - T_c)^{-\gamma}, \tag{22.2}$$

with $\gamma \approx 0.33$. This so-called *critical index* plays a fundamental role in the theory of phase transitions, describing the temperature behavior of thermodynamic values in a neighborhood of the critical point.

The obtained results may be reproduced with a reasonable accuracy within a *mean-field approximation*. In this approach, spins are considered independent while the interaction with neighbors is described by an average field, so that

$$E = -J\,q \sum_i s_i \langle s \rangle, \tag{22.3}$$

where q is the number of the nearest neighbors and $\langle s \rangle$ is the average spin. Since $s = \pm 1$ the Gibbs average (9.12) contains only the two terms

$$\langle s \rangle = f_{+1} - f_{-1} = \frac{e^{\beta q J \langle s \rangle} - e^{-\beta q J \langle s \rangle}}{e^{\beta q J \langle s \rangle} + e^{-\beta q J \langle s \rangle}} = \tanh(\beta\, q\, J\, \langle s \rangle). \tag{22.4}$$

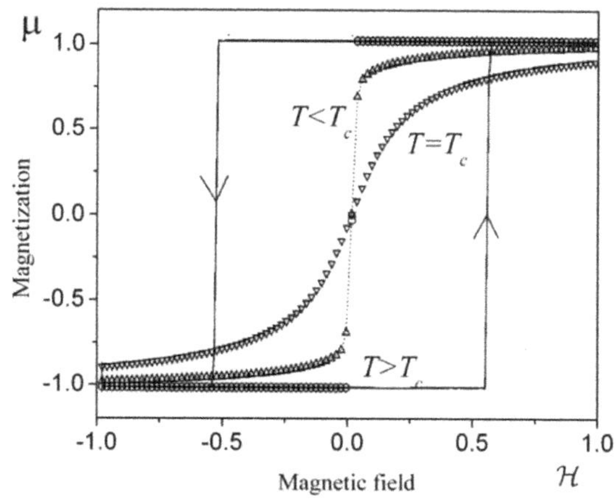

Figure 22.2. The dependence of the magnetization on the magnetic field at three temperatures: $kT = 1.0$, circles; $kT = kTs = 2.3$, hexagonals; $kT = 3.0$, inverted hexagonals.

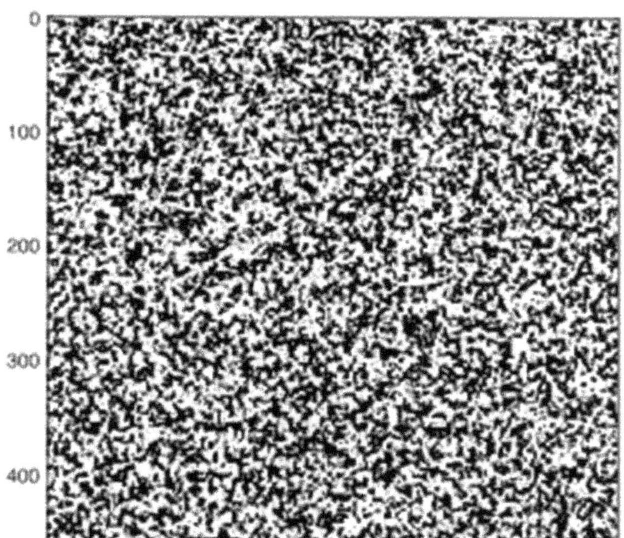

Figure 22.3. Quenching of an Ising system on a square lattice (500×500) with inverse temperature $\beta = 10$. This file is licensed under the Creative Commons Attribution-Share Alike 4.0 International license (https://creativecommons.org/licenses/by-sa/4.0). Animation available at http://iopscience.iop.org/book/978-0-7503-1341-4.

For parameters q and J, such that $qJ = 1$, the solution of this transcendental equation with respect to $\langle s \rangle$ (the points in figure 22.1) lies close to the magnetization curve calculated by the MK method. The condition $\beta q J = 1$ determines the critical temperature $kT_c = qJ = 1$ and a critical index, which, as can be seen, is $\gamma = 1/2$.

There is a deep analogy between the structural phase transitions and those observed within the Ising model and its modifications. To clarify it, we supplement the model considered above with the interaction of spins with a magnetic field $\mathcal{H}$:

$$E = -J\sum_{i,j} s_i s_j - \mathcal{H}\sum_i s_i, \tag{22.5}$$

where i, j run from 1 to N. The number of neighbors of each spin is defined by the lattice structure and space dimension of such a system.

Comparing the behavior of the magnetization and thermal capacity as functions of temperature, it is possible to reveal an analogy to a liquid–gas system. It is reached by the compliance $H \to P$ and $m \to V$. This is clearly seen in the calculated dependence of magnetization on a magnetic field for different temperatures (figure 22.2; see appendix A.10). The discontinuity in the magnetization seen in the figure for the ordered phase is the full analog of a discontinuity in density in the liquid–gas system (cf figure 21.2).

Within the Ising model it is easier than anywhere else to see how the free energy depends on the order parameter—the magnetization. Since entropy is defined as a logarithm of the number of states which is equal to the number of combinations from N on N_+: $C_N^{N_+}$ with $N_+ = N(1 + m)/2$, we obtain

$$S = k\ln\left(C_N^{N(1+m)/2}\right). \tag{22.6}$$

After insignificant simplifications, this expression reduces to

$$s = \frac{S}{kN} = \ln 2 - \frac{1}{2}(1 + m)\ln(1 + m) - \frac{1}{2}(1 - m)\ln(1 - m). \tag{22.7}$$

Calculation of energy

$$e = E/N = -\frac{J\sum_i m^2}{N} = -\frac{1}{2}qJm^2. \tag{22.8}$$

gives the free energy of equation (11.8) per particle

$$\begin{aligned} f = e - Ts = -\frac{1}{2}qJm^2 + \\ +\frac{1}{2}T\left[(1 + m)\ln(1 + m) + \frac{1}{2}(1 - m)\ln(1 - m)\right] - T\ln 2. \end{aligned} \tag{22.9}$$

The behavior of this function (without the last term) is shown in figure 22.4. To make all features of this curve more understandable expand the free energy of powers of m and $T - T_c$ near $T = T_c$:

$$f(m, T) = -T\ln 2 + \frac{1}{2}(T - T_c)m^2 + \frac{T}{12}m^4. \tag{22.10}$$

For $T > T_c$ the free energy has a single minimum corresponding to the disordered phase with $m = 0$. If we gradually reduce the temperature, i.e., increase the value of Jq/kT, the shape of the free energy around $m = 0$ gradually becomes flatter and

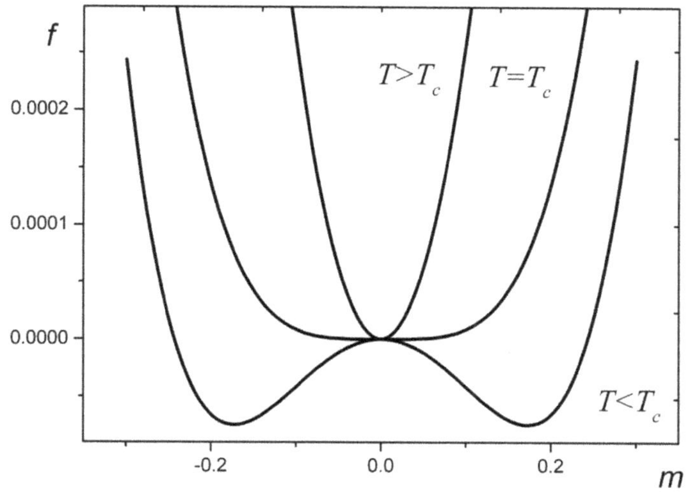

Figure 22.4. The dependence of the free energy on the magnetization. The upper curve: $T/qJ = 1.05$; middle: $T/qJ = 1.0$ and bottom: $T/qJ = 0.99$.

flatter. At a critical value $Jq/kT = 1$, shown in figure 22.4, the minimum becomes so flat that the second derivative goes to zero. At this point, the curve no longer looks like a parabola; instead, it looks like a fourth-order function. As we reduce the temperature further, the single minimum at $m = 0$ splits up into two minima at small positive and negative values of m, as shown in figure 22.4. As the temperature continues to decrease, the minima move outward and eventually approach ± 1. At a low temperature with $T < T_c$ there are two minima corresponding to the two different directions of total magnet moment $\pm m$. Because there is no applied field, the system has no preference about whether the magnetic order should point up or down. Hence, it randomly chooses one of the minima, with positive or negative m. This low-temperature state with spontaneous (not induced by field) magnetic order corresponds to the ferromagnetic phase. And, finally, the merging of both minima to a point $m = 0$ happens at a point T_c. As m changes continuously near the critical point, we are dealing here with a *second order phase transition.*

The high-temperature paramagnetic phase has a symmetry between up and down, with no preference for either direction. Hence this state has more symmetry. In the low-temperature ferromagnetic phase, the symmetry is broken and the system randomly goes one way or the other. This random selection is called *spontaneous symmetry breaking*.

Problem 22.1. Express the energy of a spin system in a large magnetic field through the absolute temperature.

Further reading

Salinas S R A 2003 Phase transitions and critical phenomena: classical theories *Introduction to Statistical Physics.* (Graduate Texts in Contemporary Physics)

IOP Publishing

Principles of Statistical Physics and Numerical Modeling

Valeriy A Ryabov

Appendix A: Details of the calculation algorithms

Appendix A.1

The numerical solution of the equation of motion of (2.1) is carried out by means of the so-called *Verlet algorithm*

$$\begin{aligned} \mathbf{r}(\Delta) &= \mathbf{r}(0) + \dot{\mathbf{r}}(0)\Delta + \frac{\mathbf{F}(0)}{m}\frac{\Delta^2}{2}, \\ \mathbf{p}(\Delta) &= \mathbf{p}(0) + (\mathbf{F}(0) + \mathbf{F}(\Delta))\frac{\Delta}{2}, \end{aligned} \tag{A.1}$$

which calculates successive values of coordinates and impulses with the time step Δ.

Unlike the usual decomposition of quantities $\mathbf{r}$ and $\dot{\mathbf{r}}$ in the Taylor series of powers Δ, Verlet's algorithm, instead of the force $\mathbf{F}(0)$, contains their arithmetic average relating also to the following step in time. Without going into detail, we note that this small replacement makes the algorithm of equation (A.1) reversible in time. In addition, it keeps a constant volume of phase space with time. These properties are inherent in exact motion equations and therefore are particularly important for the reliability of simulations in statistical mechanics.

Numerical implementation of the solution of the system of motion equations by Verlet's method involves the following steps.

1. Set the initial coordinates and impulses of the particle $\mathbf{r}(0)$, $m\dot{\mathbf{r}}(0)$.
2. Calculate the new position of the particle in the following step in time Δ:

$$\mathbf{r}(\Delta) = \mathbf{r}(0) + \mathbf{v}(0)\Delta + \frac{1}{m}\mathbf{F}(0)\frac{\Delta^2}{2}. \tag{A.2}$$

3. Calculate the force $\mathbf{F}(\mathbf{r}(\Delta))$ with respect to the new position on the following step.

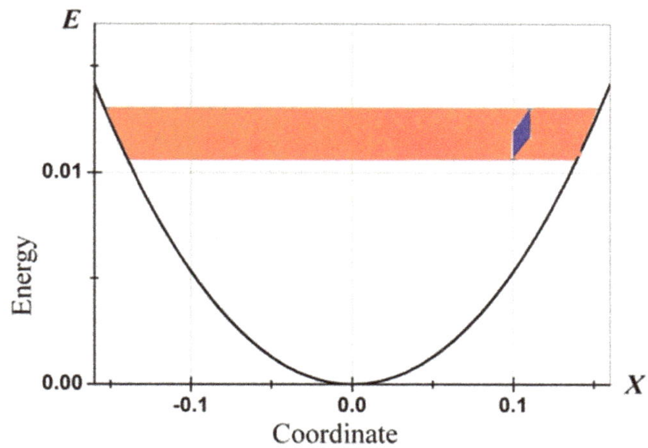

Figure A.1. The distribution of the energy for a system of 1D independent oscillators. The initial points lie in the blue box.

4. The new value of the velocity is calculated

$$v(\Delta) = v(0) + \frac{1}{m}(\mathbf{F}(0) + \mathbf{F}(\Delta))\frac{\Delta}{2}. \tag{A.3}$$

5. The whole procedure is repeated, starting from step 2, checking the conservation of the full energy.

The equations of (A.2) and (A.3) were solved for a system consisting of N = 62 500 independent 1D oscillators, whose initial phase coordinates uniformly fill the square $x_i = 0.1 \div 0.11$; $p_i = 0.1 \div 0.11$; $i = 1 \div N$ with a step $\Delta x = \Delta p = 0.0004$. The initial energy of a particle in each point $E = p^2/2 + U(x)$ with a potential $U^0(x) = x^2/2 + 2x^3/3$ was fixed and tagged by color in eight groups from $E_{\min} = 0.01\,074$ to $E_{\max} = 0.01\,306$. Figure A.1 shows the energy distribution of particles (red) in a potential well $U(x)$. The blue rhombus marks the region of initial points. For a large interval of time, all of the accessible area marked in red is almost evenly filled.

How the total energy of the system is distributed between kinetic and potential parts in the course of evolution is shown in figure 2.4. It is seen that after a very long period of time, the share of both kinetic and potential energy is close to 1/2, which requires *the virial theorem*, known from classical mechanics. The values K and U become close to the averages on time and practically do not change in times $t \sim 10^3$ when randomization of the phases of the movement occurs.

Appendix A.2

The interaction potential $U(\mathbf{r})$ shown in figure 3.2 is periodically repeated in the XY plane with a period equal to unity in both directions:

$$U^0(r) = \cos^2(\pi r). \tag{A.4}$$

The typical trajectory $ABB'CC'$ (circles) of motion is presented in figure A.2.

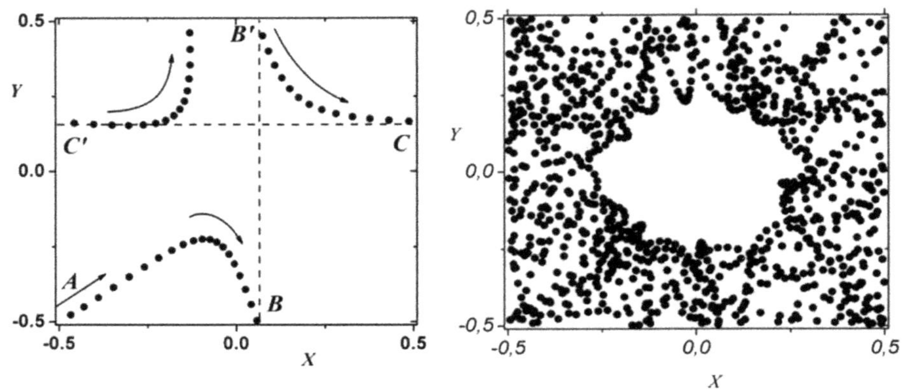

Figure A.2. The trajectory of a particle in the cell per single scattering center. Right: the same particle trajectory for a longer time

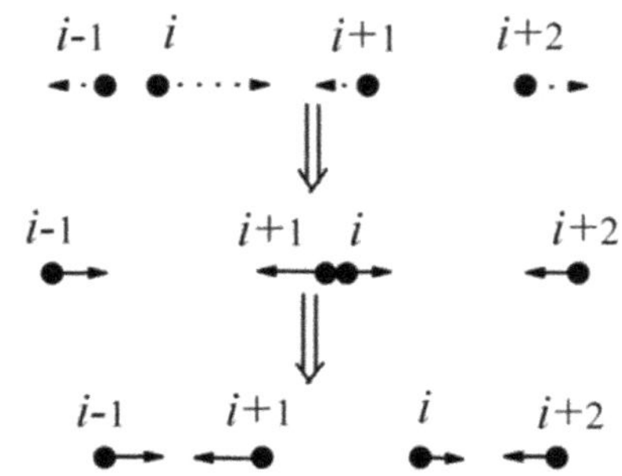

Figure A.3. A scheme of impulse exchange in the collision of pairs of particles.

It is reduced by means of periodic boundary conditions in a single cell. The direction of movement is indicated by the arrows. The calculation was made by means of the integrator of Verlet and potential of equation (A.4) with the initial coordinate $\mathbf{r}(0) = (-0.5, -0.5)$, the initial momentum $p(0) = (0.6, 0.4)$ and $m = 1$. The step in Verlet's algorithm was $\Delta = 0.001$.

The dotted line shows the transfer of a particle on reaching the cell border (the points of B and C) at equivalent points on the other side of a cell (B' and C', respectively) with the same value of velocity for the trajectory represented in the 3D option in figure 3.2. There and in figure 3.2, the trajectory points separated by an equal period $\Delta t = 0.1$ are marked by circles. The right part of figure A.2 shows the same trajectory, but for a longer time $t = 50$. It is seen that the spatial density of points is still far from uniform (cf figure 3.1). The empty area in the center of the picture is the classically inaccessible region, for which the condition $E_0 < U(r)$ is not satisfied.

Appendix A.3

A scheme of impulse exchange in the collision of couples of particles is shown in figure A.3. In the upper part of the figure, an example of positions and momenta (arrows) before the collision is shown. The middle shows the moment of the collision and the bottom of the figure the particles after the collision.

From the variety of all the options corresponding to this scheme, we choose the simplest option in which the relationship between the initial p_{i-1} p_i p_{i+1}, p_{i+2} and final p'_{i-1} p'_i p'_{i+1}, p'_{i+2} momenta at collision point x_{i-1} x_i is given by

$$\begin{aligned} p'_{i-1} &= \frac{1}{2}(p_{i-1} + p_i - p_{i+1} + p_{i+2}), \\ p'_i &= \frac{1}{2}(p_{i-1} + p_i + p_{i+1} - p_{i+2}), \\ p'_{i+1} &= \frac{1}{2}(p_{i-1} - p_i + p_{i+1} + p_{i+2}), \\ p'_{i+2} &= \frac{1}{2}(-p_{i-1} + p_i + p_{i+1} + p_{i+2}). \end{aligned} \tag{A.5}$$

A direct check of this ratio shows that a collision of this kind preserves the total energy and an impulse.

The program of collisional dynamics of a gas of free particles in 1D space consists of the following steps.

1. Set the initial coordinates $x_i = x_i^0 = i\,d$.
2. Simulate the initial impulses ($m = 1$),

$$p_i = \delta(2\xi - 1), \tag{A.6}$$

   where ξ is a random number sampled from the uniform distribution on the interval (0,1).
3. Calculate all possible times before collision

$$t_i = \left| \frac{x_{i+1} - x_i}{t_{i+1} - t_i} \right|, \quad i = 1, 2, \ldots, N. \tag{A.7}$$

4. The minimum duration is selected $t_{\min} = t_{i=m}$.
5. According to equation (1.5), calculate the new values of velocities after the collision v_{m-1}, v_m, v_{m+1} and v_{m+2}.
6. New coordinates of all particles are established at the time of collision: $x_i = x_i + v_i t_m$.
7. The procedure is repeated in step 3.

The momentum distribution was determined by the frequency count of values p seen in figure 4.6. The red curve is the normal distribution with dispersion $\overline{p^2} = \delta^2/3$ corresponding to Maxwell's distribution for free particles. The blue dashed line indicates the initial distribution of the momenta.

The following numerical experiment concerns the process of establishing equilibrium in a system consisting of two volumes of gas at different temperatures, T_1 and $T_2 = 0.01\ T_1$, which are brought into contact. For one volume the initial spread in uniform distribution was $\delta = 1.0$ for another $\delta = 0.1$. The procedure of reduction of each volume of $N = 500$ particles to equilibrium (for this purpose dynamics was initiated and traced up to $t \sim t_{eq}$) is identical to that described above. Further, after association of both volumes in one point of contact $x = 500$, the dynamics proceeded. At the time $t = 10^5$ velocities were fixed (the right-hand side of figure 4.6) to determine their distribution with a form of normal distribution.

Appendix A.4

The 1D version of the Verlet algorithm looks particularly simple.

1. Set the initial coordinates of atoms in the chain

$$x_i = x_i^0 = i\,d + 0.5\,\delta\,(2\xi - 1), \quad i = 1, 2, .., N, \tag{A.8}$$

   where ξ is a random number sampled from the uniform distribution on the interval (0, 1). All initial velocities are chosen equal to zero.

2. Calculate the new coordinates in the following step in time Δ,

$$x_i(\Delta) = x_i(0) + \dot{x}_i(0)\,\Delta + \frac{1}{m}F_i(0)\frac{\Delta^2}{2}. \tag{A.9}$$

3. With these coordinates, the force $F_i(\Delta)$ is calculated.
4. Calculate the value of the velocity

$$\dot{x}_i(\Delta) = \dot{x}_i(0) + \frac{1}{m}(F_i(0) + F_i(\Delta))\frac{\Delta}{2}. \tag{A.10}$$

5. The whole procedure repeats, starting from step 2.

As an example of a numerical solution of the system of the equations of motion, consider the dynamics of a chain of $N = 2000$ atoms, whose initial coordinates are specified with $\delta = 0.1$. The pair potential for two neighboring atoms of equation (1.6) is given in the form

$$U(x) = \Delta^2/2 - 2\Delta^3/3, \quad \Delta = x_i - x_{i-1} + 1. \tag{A.11}$$

The choice of this potential is not casual. For a potential of parabolic type, the dynamics is equivalent to the dynamics of the ensemble with independent oscillators at which the ergodic transition does not come in the space of any dimension.

But, even taking into account an anharmonicity, not all degrees of freedom in 1D and even in 2D ensembles participate in the transition to chaotic behavior. Already in the very first MD calculations for modeling the dynamics of a 1D chain, it was found that, together with close to ergodic behavior, in 1D dynamics an instability with respect to undamped long-wave excitations (solitons) takes place. Their emergence is simply due to anharmonicity of the pair potential. This phenomenon was certainly present in these calculations and is described in the following.

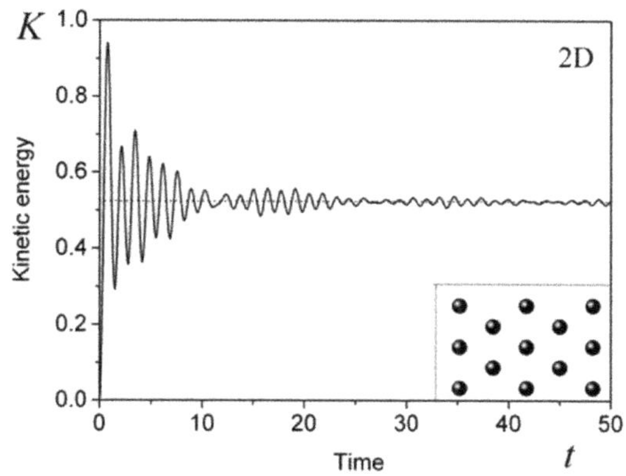

Figure A.4. Dynamics of a 2D lattice. The inset shows the hexagonal lattice.

With the growth of the number of particles at some time, the uncontrolled growth of vibration amplitude occurs. For this reason, in the 1D case the oscillation amplitude of kinetic energy is not damping with growth of the number of particles $\sim 1/\sqrt{N}$, which would indicate the equilibrium approach. This mode of instability depends on the number of particles and, for the potential used here, arises at $N > 2000$.

Another feature of chain 1D dynamics connected to the existence of solitons is the change of the ratio between the number of degrees of freedom participating in the collective movement like a soliton, to the total number depending on the number of particles N, which changes the mean value of the kinetic energy. The form of the anharmonic term in the expression for the pair potential of equation (1.11) minimizes the mentioned peculiarity of dynamics.

In conclusion, we note that for the majority of physically significant systems it is not possible to prove the existence of ergodicity. But this circumstance does not prevent us relying, however, on purely dynamic numerical calculations of average values, remembering that at any moment they can be compared with the averages on the corresponding equilibrium distributions.

For a 2D case, the instability discussed above is practically absent, but fluctuations of kinetic energy, although small, nevertheless do not yet converge to fluctuations $\sim 1/\sqrt{N}$. As an example, results of the simulation of the dynamics of a trigonal 2D lattice whose unit cell is an equilateral hexagonal (shown in the inset) are given in figure A.4. The initial coordinates of $N = 800$ particles are equal to $x_{i\alpha} = x_{i\alpha}^0 + 0.5\,\delta\,(2\xi_\alpha - 1)\, i = 1 \div N$, where $x_{i\alpha}^0$ are the coordinates of the cells of the ideal lattice. It is convenient to present the latter as a superposition of two rectangular lattices[1]

$$\mathbf{r}^1 = i_1\mathbf{a}_1 + i_2\mathbf{a}_2, \quad i_1, i_2 = 1 \div I, \quad N = I^2 \tag{A.12}$$

[1] Although these positions could be given by a single lattice (the so-called Brave lattice), the convenience in this case is due to the formation of a rectangular MD cell for which it is particularly easy to introduce PBCs.

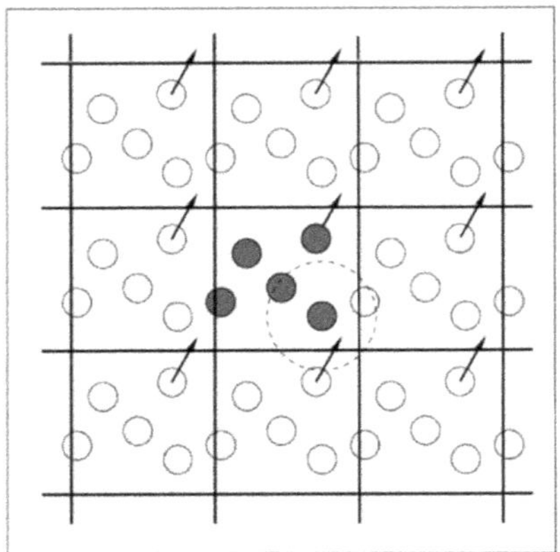

Figure A.5. The periodic boundary conditions.

and

$$\mathbf{r}^2 = \mathbf{r}^1 + 0.5\,\mathbf{a}_1 + 0.5\,\mathbf{a}_2, \tag{A.13}$$

with basis vectors $\mathbf{a}_1 = (d_x, 0)$ and $\mathbf{a}_2 = (0, d_y)$, where $d_x = 1$ and $d_y = \sqrt{3}$. The random numbers ξ_α evenly distributed in an interval (0, 1) were set just as in the linear chain.

For calculations, the pair potential of interaction was used:

$$U(r) = \begin{cases} 0.5(r-1)^2/2 - 2.5(r-1)^3/3 - U_m, & r < r_m \\ 0, & r > r_m \end{cases}, \tag{A.14}$$

where the values $U_m = 0.02666667$ and cut-off $r_m = 1.4$ provide vanishing of this potential and its derivative for the distance $r > r_m$. As is seen from the inset in figure A.4, this pair potential takes into account the interaction only between the six nearest neighbors.

Numerical realization of the Verlet algorithm in the 2D case differs from the scheme given by equations (1.2) and (1.3), through introduction of one additional line for the y-component of the coordinates and velocities, identical to those for an axis X.

Use of PBCs for a solution of the motion equations in a 2D lattice is illustrated in figure A.5. The central cell containing N particles (marked in gray) is repeated periodically throughout the space. In its identical repetitions, velocities of analogous particles are in the same direction (illustrated by an arrow at one of the particles), and coordinates differ by a translation vector. If a truncation area of potential (dotted line) of one of the particles is beyond the cell boundary, it interacts with atoms of the neighboring cells as shown in figure A.5.

The generalization for the 3D case is completely analogous. Figure A.6 shows the results for a body-centered cubic (*BCC*) lattice with the lattice parameter $d_0 = 1.0$. It

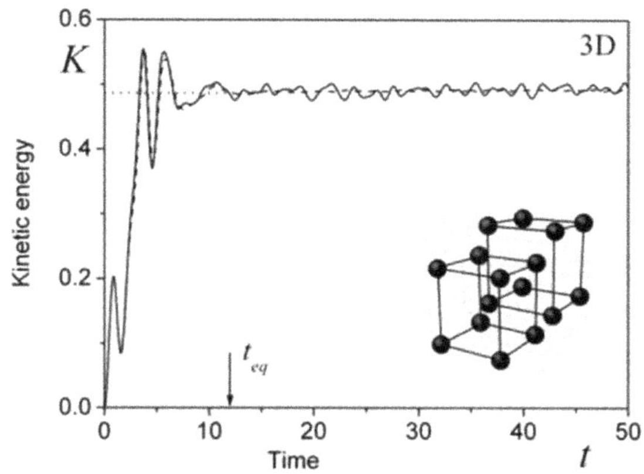

Figure A.6. Dynamics of 3D lattice. The type of lattice, *BCC*, is shown in the inset.

consists of two simple cubic lattices, inserted one into another, as indicated in the inset. All details of the calculations and the initial conditions are the same, as in the 2D lattice. The *Lennard-Jones* pair potential

$$U_{LJ}(r) = \varepsilon\left[\left(\frac{\sigma}{r}\right)^{12} - 2\left(\frac{\sigma}{r}\right)^{6}\right] \tag{A.15}$$

was used in calculations with the parameters $\varepsilon = 1.0$, $\sigma = \sqrt{3}/2d_0$ (equal to the distance to the eight nearest neighbors) and cut-off value $r_c = 0.9d_0$ ($U_{LJ}(r > r_c) = 0$).

The solid line in figure A.6 shows the result of the simulation for $N = 2000$ particles and the dotted line for $N = 128\,000$ particles. Visible reduction in the spread of the order of ~10 times shows almost complete ergodic transition in this case.

Appendix A.5

Modeling of the dynamics of a chain of $N = 2000$ atoms of mass $m = 1$, whose initial coordinates are spread within $\delta_{max} = 0.07 \div 0.11$, and the period of the chain $d - 1 = -0.005 \div 0.02$ was performed with the pair potential of the interaction $U(x)$ of equation (1.14). This potential is asymmetric with respect to $\Delta = x_i - x_{i-1}$. Only in this case, as can be seen from the expression of equation (7.5) for pressure, does it have no minimum near the equilibrium value $d = 1, P(T = 0) = 0$. Otherwise, the optimization procedure would make no sense.

Appendix A.6

Imagine the thermostat as an ideal gas of particles of m_{th} obeying the Maxwell distribution with $m_{th}\langle v_{th}^2\rangle = kT$. The interaction of a 1D oscillator of $M = 1$ with a thermostat is described as a sequence of collisions accompanied by momentum transfer: $m(\mathbf{v}'-\mathbf{v}) = m_{th}\mathbf{v}_{th}$. The calculation of the oscillator trajectory suggests the change of velocity

$$\mathrm{v}' = \frac{(1 - m_{th})\mathrm{v} + 2m_{th}\mathrm{v}_{th}}{1 + m_{th}} \tag{A.16}$$

in a time interval Δ. In each collision, the change of the oscillator energy is determined by the random values of $\mathrm{v}_{th} = \sqrt{kT/m_{th}}\,\xi$ where ξ is sampled via to the normal distribution. The initial location of the oscillator is on the bottom of the potential well $U(x) = 0.5x^2$. The animation represents the results of calculation with parameters $m = 1$, $m_{th} = 0.001$, $t = 300$, $\Delta/t = K$ and $K = 43$. The seemingly strange magnitude of K was chosen to avoid possible resonances.

Appendix A.7

Generalization of the Verlet algorithm to the case of the isothermal (NVT) ensemble is complicated by the calculation of velocity in the second step in equation (10.4). Although the values of particle velocities are available only at the end of the procedure, nothing prevents us, however, from determining it from the system of the resulting algebraic equations. The sequence of operations is as follows.

1. Calculate the coordinates of particles and the coordinate of the thermostat

$$\xi(\Delta) = \xi(0) + \dot{\xi}(0)\Delta + \frac{f_\xi(0)}{Q}\frac{\Delta^2}{2}, \tag{A.17}$$

where

$$f_\xi = \frac{2}{N}K - kT \tag{A.18}$$

designates the 'force' acting on the thermostat.

2. Solve (for example, by iterations) the system of the algebraic equations

$$\dot{x}_i(\Delta)\left(1 + \frac{1}{2}\dot{\xi}(\Delta)\right) = \dot{x}_i(0) + \frac{1}{m}(F_i(0) + F_i(\Delta))\frac{\Delta}{2} - \frac{1}{2}\dot{x}_i(0)\dot{\xi}(0). \tag{A.19}$$

To improve the accuracy of this procedure, the starting value of the velocity of the thermostat is set equal to

$$\dot{\xi}(\Delta) = \dot{\xi}(-\Delta) + 2\frac{f_\xi(0)}{Q}\Delta. \tag{A.20}$$

After each iteration, the obtained values of $\dot{x}_i(\Delta)$ are used for the determination of kinetic energy and, consequently, $F_\xi(\Delta)$.

3. Calculate the thermostat velocity

$$\dot{\xi}(\Delta) = \dot{\xi}(0) + \frac{1}{Q}\left[f_\xi(0) + f_\xi(\Delta)\right]\frac{\Delta}{2}. \tag{A.21}$$

The process of iterations, starting with step 2, repeats until the velocities of particles practically cease to change (five iterations are usually enough for a relative change of $\dot{x}_i(\Delta)$ not to exceed 0.01%).
4. The whole procedure repeats from step 1.

This procedure was used for calculation of the time evolution of the 2D lattice described in chapter 5. The initial displacements of atoms are sampled with $\delta = 0.01$. The temperature $kT = 0.001$ was maintained during evolution in the two runs with $Q = 0.1$ (blue curve in figure 10.1) and $Q = 0.01$ (red curve).

Appendix A.8

We consider here the *FCC* lattice of $N = 4000$ atoms with the period $d_0 = 1.0$. Unlike the *BCC* lattice (see appendix A.5) it consists of four simple cubic lattices, shown in the inset of figure A.6. The initial velocity variance corresponds to the temperature $kT = 3.0$. The parameters of the Lennard-Jones pair are equal to (1,1,1.0,2.5)[2]. The constant temperature ensemble was used to maitain $kT = 3.0$ in a series of calculations with respect to lattice constant increment $d/d_0 - 1$ displayed on axis X of figure 11.1. It should be noted that the minimum of free energy for this system is not achievable since the full energy in all simulations is positive. This means that the final point of evolution of such a system is a gaseous state (this conclusion will be confirmed below by a simulation within an isobaric–isothermal ensemble).

Appendix A.9

It is convenient to present the equations of motion in terms of the angular variables $\theta_i = x_i/\lambda$ with the help of equation (13.3):

$$\ddot{\theta}_i = \frac{1}{d}\left(\frac{F_i}{m} - 2\frac{\dot{\lambda}}{\lambda}\dot{\theta}_i\right) = \frac{F_i^*}{m\lambda},$$
$$M_B\ddot{\lambda} = f^{\lambda} - P^{\text{ext}}d_0 = \widehat{f}^{\lambda} \tag{A.22}$$

where the force $F_i^* = F_i - 2\dot{\lambda}m\dot{x}_i/\lambda$.

Now, applying Verlet's algorithm of (1.1) to these equations, and then returning to the variables x and $\dot{x}$, we obtain the required integrator for coordinates

$$\lambda(\Delta) = \lambda(0) + \dot{\lambda}(0)\Delta + \frac{\widehat{f}^{\lambda}}{M_B}\frac{\Delta^2}{2},$$
$$x_i(\Delta) = \frac{\lambda(\Delta)}{\lambda(0)}\left(x_i(0) + \frac{p_i(0)}{m}\Delta + \frac{F_i^*(0)}{m}\frac{\Delta^2}{2}\right) \tag{A.23}$$

[2] These calculations could be carried out with the help of LAMMPS, which was used for the examples.

and velocities

$$\dot{x}_i(\Delta) = \frac{\lambda(\Delta)}{\lambda(0)}(\dot{x}_i(0) + \frac{F_i^*(0)}{m}\frac{\Delta}{2}) + \frac{F_i^*(\Delta)}{m})\frac{\Delta}{2}$$
$$\dot{\lambda}(\Delta) = \dot{\lambda}(0) + \frac{1}{M_B}(\widehat{f}^{\lambda}(0) + \widehat{f}^{\lambda}(\Delta))\frac{\Delta}{2}, \quad \text{(A.24)}$$

The force $F_i^*(\Delta)$ in the first equation depends on a solution for $\dot{\lambda}(\Delta)$ in the second equation and, consequently, requires knowledge of the velocities $\dot{x}_i(\Delta)$ comprising the force $F_R(\Delta)$. Therefore, the system of equations of (A.24) can be solved self-consistently by iterations. With $\Delta = 0.001$ it is enough to execute the operations in equation (A.24) several times.

These equations were solved for the model described in the text. The calculation for the chain of $N = 2000$ atoms of mass $m = 1$, under pressure/tension $P = 0.1$, was carried out with barostat mass $M_B = 0.3$ and initial spread of coordinates $\delta_{\max} = 0.01$ figure 13.1. The dependence of dynamics on the magnitude of barostat mass is illustrated in figure A.7. Regardless of this mass, the instantaneous temperature of the ensemble tends to the same value (a yellow dotted line), while the kinetic energy of the barostat vanishes, although in a different time, to achieve equilibrium. Recall that the dynamics of isobaric and isotension 1D ensembles in the 1D case do not coincide.

The system of equations of motion for an *isotension–isoenthalpic ensemble* in the 2D case does not differ from the equations of (A.24) for the 1D case, except for the replacement of coordinate x_i with the component $x_{i,\,\alpha}$ and $f^{\lambda} \to f_{\alpha}^{\lambda}$ of equation (7.6) relating to the axes of biaxial deformation X_α:

$$x_{i\alpha} = \lambda_\alpha \theta_{i\alpha},$$
$$\dot{x}_{i\alpha} = \dot{\lambda}_\alpha \theta_{i\alpha} + \lambda_\alpha \dot{\theta}_{i\alpha},$$
$$\ddot{\theta}_{i\alpha} = \frac{1}{d}(\frac{F_{i\alpha}}{m} - 2\frac{\dot{\lambda}_\alpha}{\lambda_\alpha}\dot{\theta}_{i\alpha}) = \frac{F_{i\alpha}^*}{m\lambda_\alpha}, \quad \text{(A.25)}$$
$$M_B \ddot{\lambda}_\alpha = f_\alpha^{\lambda} - f_\alpha^{\text{ext}} = \widehat{f}_\alpha^{\lambda}.$$

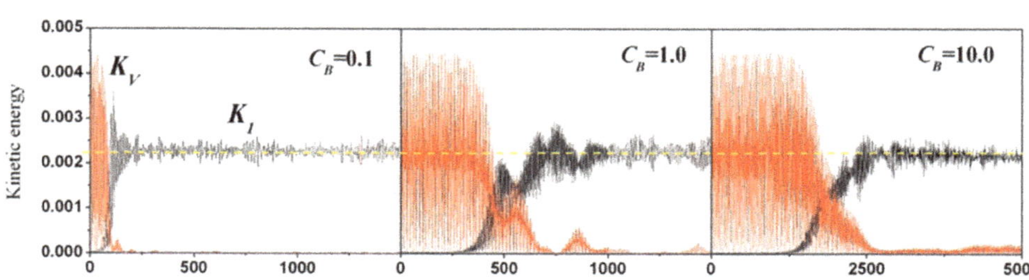

Figure A.7. The dependence of the dynamics of an isoenthalpic ensemble on barostat mass.

The volume in the run varies as $V = V_0\lambda_x\lambda_y$. The equations of motion of (A.25) conserve the instantaneous enthalpy

$$\Phi = \frac{1}{2}\sum_{i=1,\alpha}^{N}\left(\frac{m\,v_{i\alpha}^2}{2} + \frac{M_B}{2}\dot{\lambda}_\alpha^2 + f_\alpha^{\text{ext}}\lambda_\alpha\right) + U, \tag{A.26}$$

where the velocity $v_{i\alpha} = \lambda_\alpha\dot{\theta}_{i\alpha}$

The solution of these motion equations for the masses $m = M_B = 1$ shown in figure A.8 relates to an ideal trigonal 2D lattice, initially deformed along the X-direction: $d_{0x} = 1.01$ and $d_{0y} = 1.0$ ($f_\alpha^{\text{ext}} = 0$). In the perpendicular direction Y, a deformation is absent and so the Verlet algorithm is sufficient here. The directions of both axes are marked by yellow dots. All other parameters have been described in appendix A.4.

The next example concerns a simulation of the isobaric–isoenthalpic ensemble for the 3D *BCC* lattice described in appendix A.5. Compared to the 2D case discussed,

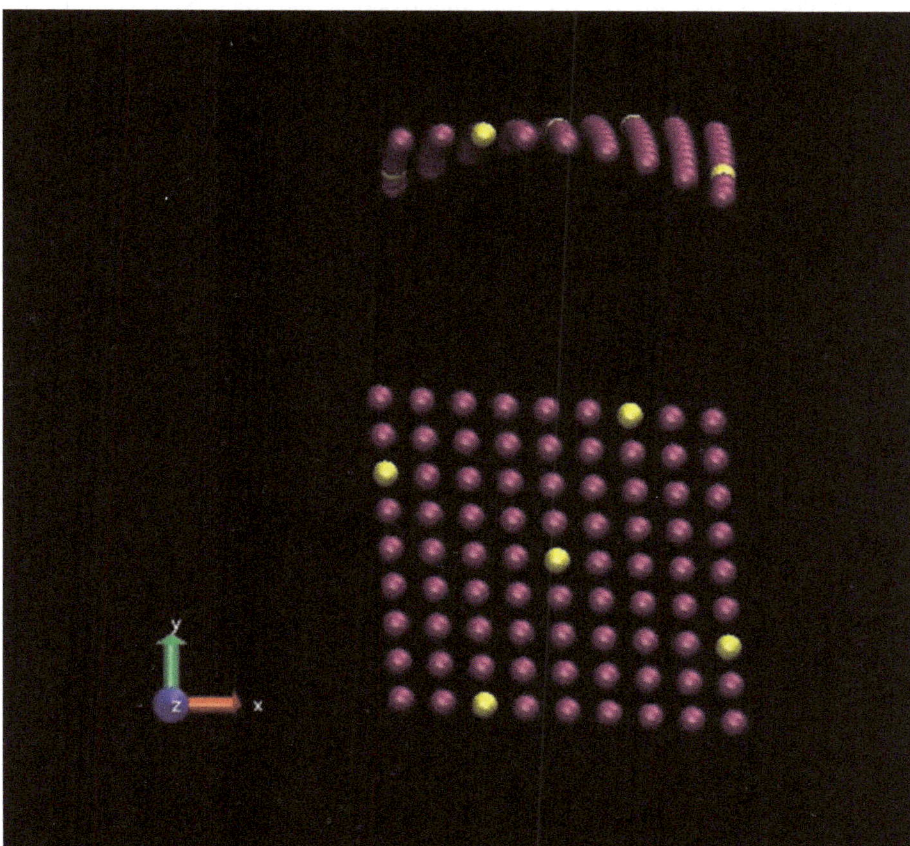

Figure A.8. Uniaxial strain along X axis. Direction of Y is parallel to cylinder axis.

the first equation of (1.22) is repeated now for all three axes. The expression of the deformation force changes to

$$\widehat{f}^{\lambda} = \frac{(P - P^{\rm ext})\upsilon}{\lambda}, \tag{A.27}$$

where $\upsilon = V/N$ is the volume per particle which depends on the expansion: $\upsilon = \upsilon_0\lambda^3$. Unlike in the 1D case where $\upsilon \to d = \lambda d_0$, the volume is now not reduced with the stretch ratio.

The 3D simulations of the isobaric 3D ensemble were made for a model similar to that used in appendix A.4. We consider the *FCC* lattice with the lattice parameter $d_0 = 1.0$ seen in the inset of figure A.9. It consists of four simple cubic lattices. If one cubic lattice is attached to the origin of coordinates, the three others are shifted along the diagonals of a face: [110], [101] and [011]. The LJ potential with the parameters $\varepsilon = 1.0$, $\sigma = \sqrt{2}/2d_0$ (equal to the distance to the eight nearest neighbors) and cut-off value $r_c = 0.9d_0$ was used in the calculation. Three magnitudes of temperature are given by the initial spread of atomic displacements with $\delta = 0.01$, 0.05 and 0.07. Pressure evolution is shown in figure A.9. To improve the convergence of pressure (as for other macroscopic parameters) to equilibrium, a special quenching-down procedure of the strains was applied. The kinetic energy of deformations vanished each time when it reached a maximum. Therefore, the system comes to the thermodynamic limit almost immediately. At each such step, this kinetic energy was subtracted from the full initial energy of the system, to ensure the accuracy of each step (conservation of energy).

As a result of simulations, the thermal expansion was also computed. It follows from the dependence $\upsilon(T)$ for a constant pressure. Of course, this dependence could be established within the microcanonical ensemble. But in this case, we would have to adjust a volume to predetermined pressure. The results are shown in figure A.10 for three magnitudes of external pressure. Almost linear dependence of the volume on temperature means that deformation probes a small energy range where the

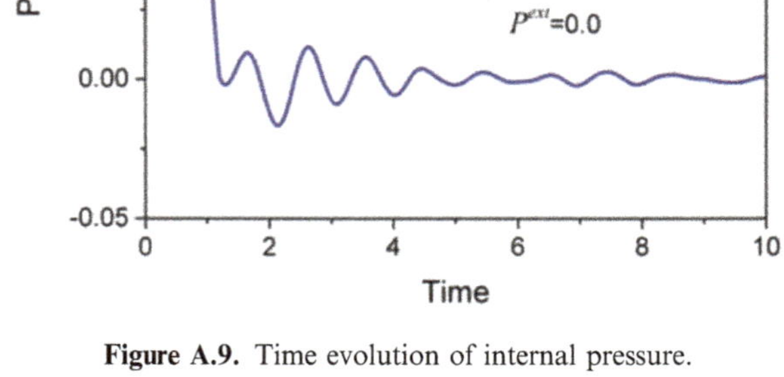

Figure A.9. Time evolution of internal pressure.

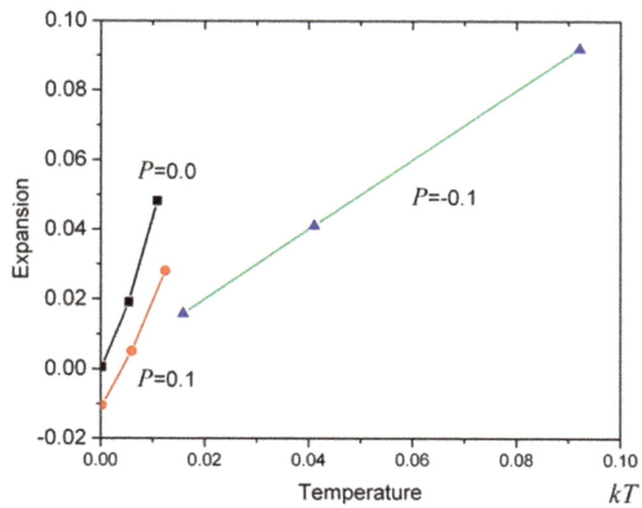

Figure A.10. Thermal expansion for different pressures.

harmonic form of the potential gives the main contribution. The coefficient of thermal expansion is given by the slope of these lines.

Appendix A.10

The implementation of a random walk on energy occurring with the probability of equation (15.12) is as follows.

1. Choose a random spin s_i in a square lattice.
2. Try to change its sign. For this purpose
   (a) calculate the energy change ΔE (for a square lattice $\Delta E = 4J$);
   (b) if $\Delta E < 0$ fix spin change;
   (c) if $\Delta E > 0$, fix spin change with probability $\exp(-\Delta E/kT)$.
3. Go back to step 1.

In the process of transitions calculate the energy of configuration e_N and entropy s_N which further is used for determination of the microcanonical temperature.

Appendix A.11

The initial velocities of the particles are chosen as

$$v_{i\alpha} = \sqrt{kT}\,\xi, \quad \alpha = x, y,$$

where ξ is a random number sampled from the normal distribution. Then these velocities are reduced to the total momentum equal to zero $\sum_{i=1}^{N} \mathbf{v}_i = 0$. The initial atomic coordinates $x_{i\alpha} = l_0\,(2\eta - 1)$ are sampled from the uniform distribution for η on the interval (0, 1). During the run of the time $\tau = 1000$, straight trajectories are tracked until a collision with the cap where the reflection of the y-component of momentum takes place. All momentum transfers $2v_y$ are added up to calculate the averaged force $F = Pl_0$ acting on the cap. This force is used in the Verlet integrator for the motion equation of a cap of mass M_B (the weight m_P is neglected here).

Then the kinetic energy rises to $E = E + \Delta E$ and the new run of momentum gain starts again.

Appendix A.12

Calculations of the potential profile and dynamics of atoms in the hexagonal lattice were made using the *Lennard-Jones potential*

$$U_{\rm LJ}(r) = 0.1\left(\frac{1}{r^{12}} - \frac{2}{r^6}\right). \tag{A.28}$$

To provide the continuous logarithmic derivative in a sewing point $r_1 = 1.2$ and in the cut-off r_m, the compound function was used

$$U(r) = \begin{cases} U_{\rm LJ}(r), & r < r_1; \\ -a\,(r - r_m)^{3/2}, & r_1 > r > r_m; \end{cases} \tag{A.29}$$

with parameters $r_m = 1.575\,529\,345$ and $a = 0.242\,3190\,472$. It ensures that at the moment when the atom moves along a jump trajectory changing the nearest neighbors, the force and potential vary in a continuous manner.

To determine potential profile near the reaction path (dashed line in figure 19.3) atom A was sequentially placed into positions belonging to the rectangle with coordinates $x \in 0,\ 1;\ y \in -0.1,\ \ 0.1$ towards the vacancy located at the point $\mathbf{r}_B = 0$. At a fixed position of atom A on the reaction path, dynamics was initiated, i.e., the equations of (24.1) were solved, starting from an ideal arrangement of all other atoms of the crystallite. However, there is one circumstance which we have not yet taken into account in this scheme. Atom A at any point $\mathbf{r}_A$ inside the rectangle acts on the surrounding atoms that, eventually, lead to displacement of the whole lattice to a minimum where this force disappears (due to PBC). To compensate this force and, thereby, to provide this atom immovability with respect to the lattice, we have to place the other atom at a point $-\mathbf{r}_A$ so that deformation fields from both defects do not overlap: $r_A \gg 1$. To this inverse *image* should be added also two other atoms at points $(x_A,\ -y_A)$ and $(-x_A,\ y_A)$ to exclude the torque and MD of the cell rotation. As a result, the trajectories of all four atoms in the images give the same potential of atom A for the fixed lattice.[3]

During the relaxation procedure, the potential energy decreases and the kinetic energy increases. As soon as the kinetic energy reaches a maximum, all momenta $\mathbf{p}_i$ are set to zero and the process repeats until the kinetic energy becomes neglible, not exceeding some preset value $K_1^{\rm min}$. Only then can it be assured that all atoms of a lattice are in equilibrium. In this calculation, it was assumed that $K_1^{\rm min} = 10^{-5}$.

The values of the potential energy obtained for the lattice consisting of $N = 799$ particles are shown in figure 19.3. The reaction trajectory is defined as a set of points

[3] In principal, this procedure can be avoided by use of the fixed boundary conditions.

of extremum in the Y-direction. The barrier height in the saddle-point on a trajectory (the *activation energy of diffusion* or *migration energy*) was $U_{\text{barrier}} \approx 0.37$.

Further, within the same MD cell, the dynamics of the NVE ensemble was initiated (see chapter 5). The jump was fixed while one of six atoms surrounding the vacancy crosses the boundary of the *Wigner–Seitz cell.* In figure 19.1, this is marked by a dashed line and represents a hexagon with its sides in the middle of the space between atoms and their nearest neighbors. From geometrical considerations the condition of atom $\mathbf{r}_A$ to be in a lattice cell $\boldsymbol{B}$ can be written in the form of a set of inequalities

$$|\, \mathbf{n}_{BB'}(\mathbf{r}_A - \mathbf{r}_B^0)| < \frac{1}{2} r_{BB'}, \tag{A.30}$$

where $\mathbf{n}_{BB'}$ is the unit vector directed along vector $\mathbf{r}_{BB'} = \mathbf{r}_{B'}^0 - \mathbf{r}_B^0$ connecting the lattice cell $\boldsymbol{B}$ with nearest cells $\boldsymbol{B}'$.

Fixing times $t_m, m = 1 \div M$ for separate jumps for the time of the run t it is possible to construct a distribution function of their duration $\tau_m = t_m - t_{m-1}$. The count frequency gives the exponential distribution with an average duration of a jump being $\tau = t/M$. In figure 19.4 it is represented as a function of the inverse temperature $\beta = 1/kT$ in the logarithmic scale. The linear dependence also verifies that this distribution is exponential.

Appendix A.13

It is convenient to describe the distinction in the degree of ordering of atoms by means of a so-called *pair correlation function* $\widehat{g}(\mathbf{r})$. It differs from the coordinate distribution function we encountered earlier by the origin of coordinates to be located in a mean position of one atom. It is connected with the probability of an arrangement of atoms at the distance $\mathbf{r}$ from each other

$$\widehat{g}(\mathbf{r}) = \frac{1}{N(N-1)} \sum_{i=1}^{N} \left\langle \sum_{j \neq i}^{N} \delta(\mathbf{r} - \mathbf{r}_{ij}(t)) \right\rangle.$$

The constant multiplier gives the number of combinations of possible pairs of indices, here i, j, and hexagonal triangular brackets denote the average over time. Integrating this function over angles and dividing by the average density of atoms ρ in a polar coordinate system $g(r) = 2\pi \widehat{g}(r) r/\rho$ (or a spherical one for a 3D case $g(r) = 4\pi \widehat{g}(r) r^2/\rho$) we obtain the *radial distribution function*. It does not depend on the direction and is defined only by the number of atoms in a ring (layer) $(r, r + dr)$. In figure A.11, these atoms are blue. In MD programs, this function is usually determined simultaneously with the procedure of counting the number of pair interactions with each atom.

The pair potential used in calculations and shown in figure 20.2 is

$$U(r) = \begin{cases} U_p(r) - 2U_p(r_1) + U_p(r_2), & r < r_1; \\ U_p(r_2) - U_p(2r_1 - r), & r_3 > r > r_1; \\ 0, & r > r_3. \end{cases}$$

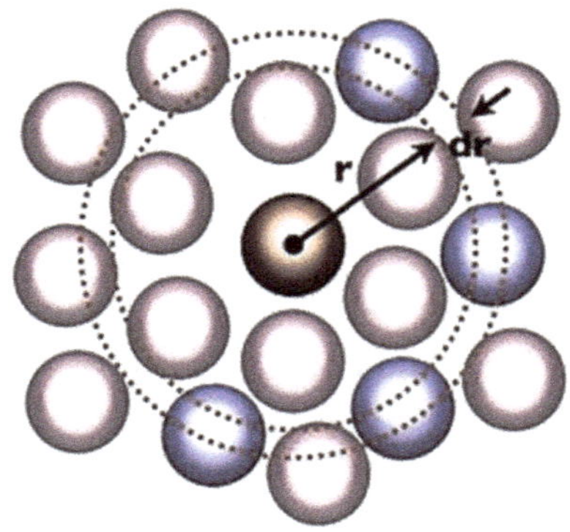

Figure A.11. Toward the definition of the radial distribution function. Atoms in the ring are in blue.

Here $U_p(r) = 5(r - 1.0)^2 - 10(r - 1.0)^3$, $r_1 = 0.5(r_2 + r_3)$, $r_2 = 1.333300$, $r_3 = \sqrt{2}$. An MD cell with a square 30 × 30 lattice was rotated through the angle $\pi/4$ to provide the dynamics of the main stretches along axes *X* and *Y*, as is shown in figure 21.3. Initial displacements of atoms according to a uniform distribution with $\delta = 0.03$ were introduced to initiate dynamics. Barostat masses $M_B = 0.1$ are used for both directions. The animation of the crystal lattice shows the time interval $t = 300$–350.

Appendix A.14

In this appendix we consider a generalization of the MD algorithm with the help of the Liouville operator and Trotter's theorem and then use it for the *NPT* simulation of the equation-of-states in the lead (Pb) system.

First consider for simplicity 1D dynamics. Let us write the left-hand side of the equation of motion of (1.6) in the symbolic-operator form, using the vector $\Gamma = \binom{x}{p}$ for one degree of freedom:

$$\frac{d}{dt}\begin{pmatrix} x \\ p \end{pmatrix} = i\widehat{L}\begin{pmatrix} x_0 \\ p_0 \end{pmatrix}. \tag{A.31}$$

Express the *Liouville operator* $i\widehat{L}$ through the sums of two operators, one of which affects only coordinates, as another momentum:

$$i\widehat{L} = \dot{x}\frac{\partial}{\partial x} + \dot{p}\frac{\partial}{\partial p} = i\widehat{L}_x + i\widehat{L}_p. \tag{A.32}$$

The result of the action of the operator $i\hat{L}_x$ can be set by solving the equation

$$\Gamma = i\widehat{L}_x\Gamma = \dot{x}\frac{\partial}{\partial x}\begin{pmatrix} x \\ p \end{pmatrix} = \begin{pmatrix} \dot{x} \\ 0 \end{pmatrix}. \tag{A.33}$$

Integration here comes to

$$\dot{\Gamma} = e^{i\widehat{L}_x t}\begin{pmatrix} x_0 \\ p_0 \end{pmatrix} = \begin{pmatrix} x_0 + p_0 t \\ p_0 \end{pmatrix}. \tag{A.34}$$

Similarly,

$$e^{i\widehat{L}_{p}t}\begin{pmatrix} x_0 \\ p_0 \end{pmatrix} = \begin{pmatrix} x_0 \\ p_0 + \dot{p}_0 t \end{pmatrix}. \tag{A.35}$$

Thus, the action of each operator translates either coordinate or momentum to their values at time t.

However, the knowledge of how both of these *propagators* work separately does not yet mean that we know how the propagator $e^{i\hat{L}t}$ itself works. It is easy to verify that the components of the operator Liouville $i\hat{L}_x$ and $i\hat{L}_p$ do not commute with each other: $i\hat{L}_x i\hat{L}_p \neq i\hat{L}_p i\hat{L}_x$. This means that

$$e^{i\widehat{L}\,t} = e^{i\widehat{L}_x t + i\widehat{L}_p t} \neq e^{i\widehat{L}_x t} e^{i\widehat{L}_p t}. \tag{A.36}$$

It might be checked directly, using the obvious operator relationship, following from the expansion of an arbitrary function in a Taylor series:

$$e^{c\frac{\partial}{\partial x}} F(x) = F(x + c). \tag{A.37}$$

The nature of the decomposition of the propagator $e^{i\hat{L}t}$ can be found using so-called Trotter decomposition for non-commuting operators, which in an approximate form (for a small period Δt) is

$$e^{i\widehat{L}_x \Delta t + i\widehat{L}_p \Delta t} \cong e^{i\widehat{L}_p \Delta t/2} e^{i\widehat{L}_x \Delta t} e^{i\widehat{L}_p \Delta t/2}. \tag{A.38}$$

This relationship is also verified by the Taylor series expansion.

Thus, the first step gives

$$e^{i\widehat{L}_p \Delta t/2}\begin{pmatrix} x(0) \\ p(0) \end{pmatrix} = \begin{pmatrix} x(0) \\ p(0) + \dot{p}(0)\frac{\Delta t}{2} \end{pmatrix}. \tag{A.39}$$

The action of the second propagator of equation (A.39) brings us to

$$e^{i\widehat{L}_x \Delta t}\begin{pmatrix} x(0) \\ p(0) + \dot{p}(0)\frac{\Delta t}{2} \end{pmatrix} = \begin{pmatrix} x(0) + p\left(\frac{\Delta t}{2}\right)\Delta t \\ p(0) + \dot{p}(0)\frac{\Delta t}{2} \end{pmatrix}. \tag{A.40}$$

Finally, the result of the operation of the first factor at equation (A.40) yields

$$\begin{aligned} e^{i\widehat{L}_p \Delta t/2}\begin{pmatrix} x(0) + p\left(\frac{\Delta t}{2}\right)\Delta t \\ p(0) + \dot{p}(0)\frac{\Delta t}{2} \end{pmatrix} &= \begin{pmatrix} x(t + \Delta t) \\ p(t + \Delta t) \end{pmatrix} \\ &= \begin{pmatrix} x(0) + \dot{p}(0)\Delta t + \dot{p}(0)\frac{\Delta t^2}{2} \\ p(0) + \dot{p}(0)\frac{\Delta t}{2} + \dot{p}(\Delta t)\frac{\Delta t}{2} \end{pmatrix}. \end{aligned} \tag{A.41}$$

Eventually, we obtain the already known Verlet's algorithm of equations (1.9) and (1.10).

Let us now define Liouville's operator for the isothermal isobaric 3D ensemble, including additional dynamic variables, that are the stretch λ associated with volume by a ratio $V = V_0\lambda^3$ and thermostat coordinate ξ. According to equation (1.27), it includes the derivatives over $\theta_{i\alpha} = x_{i\alpha}/\lambda$, ξ and the velocities $\dot{\lambda}$, $\dot{\theta}_\alpha$ of equation (30.1) and $\dot{\xi}$ of equation (14.4):

$$\begin{aligned} i\widehat{L} &= i\widehat{L_x} + i\widehat{L_p}, \\ i\widehat{L_x} &= \dot{\theta}_{i\alpha}\frac{\partial}{\partial\theta_{i\alpha}} + \dot{\lambda}\frac{\partial}{\partial\lambda} + \dot{\xi}\frac{\partial}{\partial\xi}, \\ i\widehat{L_p} &= \frac{F^*_{i\alpha}}{m\lambda}\frac{\partial}{\partial\dot{\theta}_{i\alpha}} + \frac{f^\lambda}{M_B}\frac{\partial}{\partial\dot{\lambda}} + \frac{f_\xi}{Q_\xi}\frac{\partial}{\partial\dot{\xi}}, \end{aligned} \tag{A.42}$$

The forces acting on the generalized coordinates according to equations (1.18) and (1.27) are equal

$$\begin{aligned} F^*_{i\alpha} &= F_{i\alpha} - 2\frac{\dot{\lambda}}{\lambda}\mathrm{v}_{i\alpha}, \\ f^\lambda &= \frac{(P - P^{ext})\upsilon}{\lambda}, \\ f_\xi &= \frac{2}{N}K - kT. \end{aligned} \tag{A.43}$$

The action of the propagator of equation (34.8) is reduced to the consecutive calculation of first velocities at the time $\Delta/2$, then the coordinates through time Δ, and again velocities through the interval $\Delta/2$. However, unlike the Verlet algorithm, the change in velocities of the particles on each time step causes a change of forces f^λ and f_ξ. Therefore, in the first term of the operator $\hat{L}_p$ of equation (A.42) we will select the operator dealing only with the velocity of the particles

$$\begin{aligned} i\widehat{L_p} &= i\widehat{L}_p^2 + i\widehat{L}_p^1, \\ i\widehat{L}_p^2 &= \frac{F_{i\alpha}}{m\lambda}\frac{\partial}{\partial\dot{\theta}_{i\alpha}}, \\ i\widehat{L}_p^1 &= -2\frac{\dot{\lambda}}{\lambda}\dot{\theta}_{i\alpha}\frac{\partial}{\partial\dot{\theta}_{i\alpha}} + \frac{f^\lambda}{M_B}\frac{\partial}{\partial\dot{\lambda}} + \frac{f_\xi}{Q}\frac{\partial}{\partial\dot{\xi}}. \end{aligned} \tag{A.44}$$

Further, we use Trotter's decomposition of equation (1.38):

$$e^{i\widehat{L}\Delta t} \cong e^{i\widehat{L}_p^1\Delta t/2}e^{i\widehat{L}_p^2\Delta t/2}e^{i\widehat{L}_x\Delta t}e^{i\widehat{L}_p^2\Delta t/2}e^{i\widehat{L}_p^1\Delta t/2}. \tag{A.45}$$

As a result, the procedure of step-by-step calculation of coordinates and momenta under the action of this propagator is as follows:

1. Given the forces $F_{i\alpha}$, f^{λ}, f_{ξ} and $\dfrac{\dot{\lambda}}{\lambda}$ at the initial time, calculate the velocities through

$$e^{i\widehat{L}_p^1\Delta t/2}: \quad \begin{aligned} \dot{\lambda} &= \dot{\lambda} + \frac{1}{M_B} f^{\lambda} \frac{\Delta t}{4}, \\ \dot{\xi} &= \dot{\xi} + \frac{1}{Q} f_{\xi} \frac{\Delta t}{4}, \\ \dot{\theta}_{i\alpha} &= \dot{\theta}_{i\alpha} - 2\frac{\dot{\lambda}}{\lambda}\dot{\theta}_{i\alpha}\frac{\Delta t}{4}. \end{aligned} \tag{A.46}$$

2. Calculate the new values of particle velocities

$$e^{i\widehat{L}_p^2\Delta t/2}: \quad \dot{\theta}_{i\alpha} = \dot{\theta}_{i\alpha} + \frac{1}{m}\frac{F_{i\alpha}}{\lambda}\frac{\Delta}{4}. \tag{A.47}$$

3. Calculate the new values of all coordinates

$$e^{i\widehat{L}_x\Delta t/2}: \quad \begin{aligned} \lambda &= \lambda + \dot{\lambda}\,\Delta, \\ \xi &= \xi + \dot{\xi}\,\Delta, \\ \theta_{i\alpha} &= \theta_{i\alpha} + \dot{\theta}_{i\alpha}\Delta. \end{aligned} \tag{A.48}$$

4. Calculate the new values of forces $F_{i\alpha}$, f^{λ} and f_{ξ}.
5. Repeat step 2.
6. Calculate the new values of kinetic energy and forces f^{λ}.
7. Repeat step 1.
8. Go back to the variables $x_{i\alpha}$ and $\mathrm{v}_{i\alpha}$.

Note that procedure of thermostatting can be applied in addition to any degree of freedom, including the barostatic one.

Figure A.12 shows an example of the evolution of a 2D hexagonal lattice to one of the two-phase states shown in figure 16.1. The values of initial parameters were equal to $kT = 0.13$, $P^{\text{ext}} = 0.02$, $V = 2.2$, $m = 1$, $M_B = 1$ and $Q = 0.1$. After establishing equilibrium, the kinetic energy K and the virial pressure P with an accuracy ~1% reproduce preset values of T and P. The equilibrium volume in calculations corresponds to the density $\rho = 0.45$ shown in figure 16.1.

Figure A.12 also shows also the evolution of the thermostat. The value $\gamma = \dot{\xi}/\xi$ at $K \to T$ vanishes, thereby turning off the effect of internal friction. The mass of thermostat Q determines the period of oscillation of kinetic energy near a predetermined temperature. Too small value for this parameter may prevent the establishment of equilibrium, and too large value will greatly extend the approach of values K to T (cf with comments for the mass of a barostat in chapter 13 and appendix A.7).

The next example refers to a 3D real substance. The ideal lattice of lead (Pb) represents an *FCC* structure with the lattice constant $d_0 = 0.49\,508$ nm. The initial

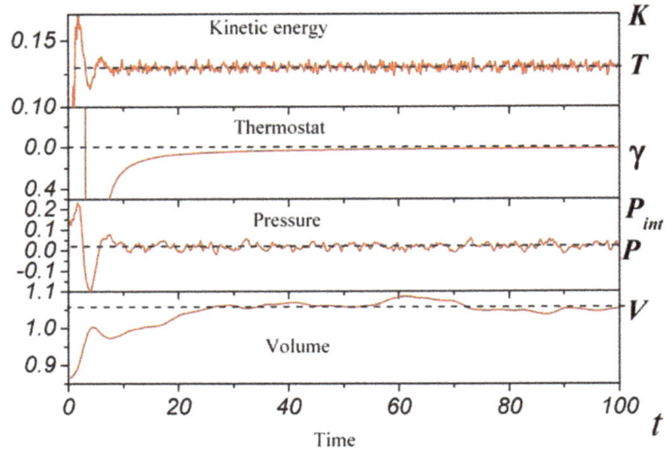

Figure A.12. Evolution of a 2D isothermal isobaric ensemble. From top to bottom: the kinetic energy; the rate of change of thermostat coordinate; pressure; volume.

arrangement of atoms near the lattice sites is sampled to initiate the dynamics and then is maintained at the predetermined value. The values of the initial parameters were $T = 250$ K, $P^{ext} = 0$, $m = 207, 2$, $M_B = 1$ and $Q = 0.1$. Each point was obtained by the procedure for the solution of the motion equation of (A.46)–(A.48) until an equilibrium is established. The second crystallization curve starts from $T \sim 2000$ K. All other features are the same as we used in earlier calculations. A lazy reader may carry out a similar calculation with the analogous ensemble and potential in the LAMMPS environment. The result will be the same, although LAMMPS uses a more advanced schema of thermostatting.

Appendix B: Solutions

Solution 1.1. The inset shows the potential and energy of the particles. When calculating the phase trajectory be aware that the accessible area for $E = 0.5$ is restricted to the values $x < 0$.

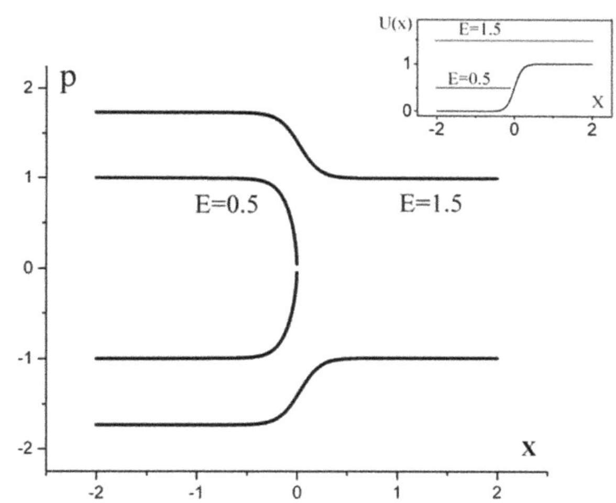

Solution 2.1. Let the particles have the coordinates x_1, x_2 and momenta p_1, p_2 before a collision and coordinates x_1', x_2' and momenta p_1', p_2' after this. Then the phase volume is conserved if

$$dx_1 dx_2 dp_1 dp_2 = dx_1' dx_2' dp_1' dp_2'.$$

Since $dx_1 dx_2 = dx_1' dx_2'$ then only $dp_1 dp_2 = dp_1' dp_2'$ are left. Substituting here the ratio between the momentum before and after collision (known from classical mechanics) one can verify that the Liouville theorem is fulfilled.

doi:10.1088/978-0-7503-1341-4ch24

Solution 2.2. According to equation (2.6)

$$f(E) = \int dx_0\, dp_0\, \delta\big(E - \mathcal{H}(x_0, p_0)\big).$$

It is more convenient to carry out the integration here in the polar coordinate system in which the product $dx_0 dp_0 = s ds d\varphi$. Since $s = \sqrt{2} \div \sqrt{2} + \Delta$ and $s\ \Delta\varphi = \Delta$ the integral goes to

$$f(E) = \int dh\ \delta(E - h),$$

where $h = \mathcal{H}(x_0, p_0)$. After integration, we obtain that $f(E)$ is the uniform distribution in the interval $E = 1 \div 1 + \sqrt{2}\,\Delta$.

Solution 3.1. The phase portrait is presented by the set of isoenergetic curves $p(x) = \sqrt{2[E_i - U^0(x)]}$, $i = 1 \div N$ with $E_i = p_{0i}^2/2 + U^0(x_i^0)$ (see appendix A.1). The difference between distributions may be seen by comparison of them within the small range shown in the inset in figure 2.3. Unlike the time distribution function of equation (2.2), all points of which are separated by a small time gap, these points of calculated distribution based on the microcanonical ensemble are merged in lines leaving their density along the axes p unchanged.

Solution 3.2. Integration in equation (3.6) comes to

$$n(x, E) = \int dp\, f(x, p) = \int dp\ \delta(E - p^2/2m - U(x)) = A\sqrt{\frac{E}{E - U(x)}}\,\vartheta(E - U(x)),$$

where A is a constant following from normalization condition

$$\int dx\ \ n(x, E) = 1.$$

It remains to carry out the summation over the initial energies $E_i = p_i^2/2 + U^0(x_i)$ for each particle

$$n_N(x) = \sum_{i=1}^{N} n(x, E_i).$$

Solution 5.1. $\langle K\rangle = \frac{3}{2}m\langle v_\alpha^2\rangle$.

Solution 6.1. The integration of the first and second moments of the velocity over the distribution of equation (6.3) gives

$$\left(\frac{1}{2}m\right)^2 \langle(v^2 - \langle v^2\rangle)^2\rangle = \frac{3}{2}(kT)^2.$$

It follows the accuracy of the determination of the microcanonical temperature $\sim 1/\sqrt{N}$.

Solution 7.1. Converting velocities $\dot{x}_i = R\,\dot{\theta}_i$ into generalized momentum $p_{\theta i} = \frac{\partial \mathscr{L}}{\partial \dot{\theta}_i} = m\,R^2\dot{\theta}_i$ in a polar coordinate system, we obtain

$$\mathscr{H} = \tfrac{1}{2}\sum\nolimits_i \frac{p_{\theta i}^2}{mR^2} + U.$$

From this follows

$$\frac{\partial \mathscr{H}}{\partial R} = -\frac{\partial \mathscr{L}}{\partial R}$$

and, consequently, the formula of equation (7.9).

Solution 7.2. Cohesion between two halves of the chain with PBCs can be calculated by removing atoms in one of two images of the finite chain. Then, the net cohesion is determined by the effect of 'missing' atoms that mimicked the excluded half. Only two forces f_{01} and f_{02} are applied to the border atoms $i = 0$ and $i = 1$ from the image atoms on the right-hand side (see figure A.1), so that $\mathsf{T} = f_{01} + 2f_{02}$. The calculation of the 1D virial stress $P = \left(\frac{1}{Nd}\right)\sum_{i=1}^{N}[f_{i,\,i-1}\cdot(x_i - x_{i-1})]$ obtains the same value $P = f_{01} + 2f_{02} = \mathsf{T}$.

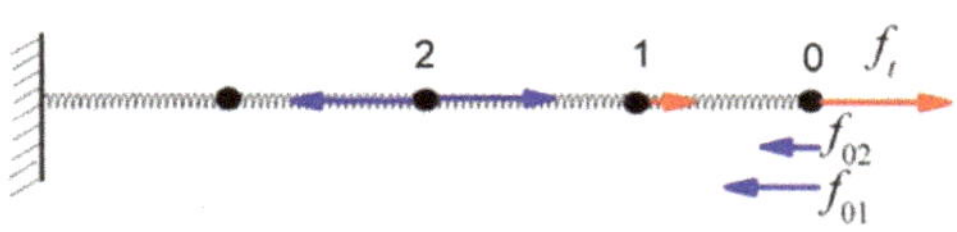

Solution 7.3. The external force $f = f_{i,\,i-1} = \kappa(x_{i,\,i-1} - d_0)$ being applied to the ends of the string of length $L_0 = Nd_0$ affects the extension to $L = Nd$ (see figure below). Since $x_{i,\,i-1} = d$ the calculation of the tension force of equation (7.5) gives $\mathsf{T} = (1/L)N\kappa(d - d_0)d = \kappa(d - d_0)$. In the 1D case, the analog of the uniaxial pressure is equal to the external force. It affects the same extension of the string as the radial force directed along the radius of the circle with PBCs.

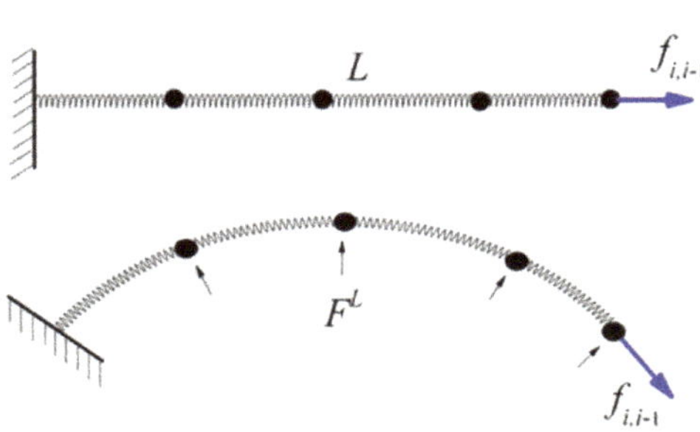

Solution 8.1. Consider the expression

$$S(\alpha) = -k\sum_{i=1}^{N}\left(p_i \ \ln p_i - \alpha p_i\right).$$

Its maximum can be found from zero derivative condition

$$\frac{\partial S}{\partial p_i} = -k\left(\ln p_i + 1 - \alpha\right) = 0.$$

It follows that p_i = const. The normalization condition for the probabilities leads to

$$S_{\max} = -k\sum_{i=1}^{N}\left(\frac{1}{N} \ \ln \frac{1}{N}\right) = k \ln N.$$

This value of the entropy corresponds to the microcanonical ensemble.

Solution 8.2. Introducing notation $x = N_1/(N_1 + N_2)$ obtain from equation (8.17)

$$\Delta S_{\text{mix}} = -(N_1 + N_2)kT[x \ln x + (1 - x)\ln(1 - x)].$$

Since both terms in square parentheses are negative, the entropy of mixing is positive and favors mixing of the pure components.

Solution 9.1. On the probability $p_i(v_i)$ are imposed conditions

$$\int_{-\infty}^{\infty} p_i(v_i)dv_i = 1, \quad \int_{-\infty}^{\infty} \frac{1}{2}mv_i^2 p_i(v_i) \ dv_i = \frac{1}{2}kT, \quad i = 1 \div N.$$

Similarly to problem 8.1, apply the method of Lagrange multipliers. For this purpose consider the function

$$f = -k\sum_{i=1}^{N}\left(\int p_i \ \ln p_i dv_i + \alpha_i' \int p_i dv_i p_i + \beta_i \int p_i \ v_i^2 dv_i\right).$$

The condition of the maximum of this function

$$\frac{\partial f}{\partial p_i} = -k\int\left(\ln p_i + \alpha_i' + \beta_i v_i^2 + 1\right)dv_i = 0$$

leads to $p_i = \exp(-\beta_i v_i^2 - \alpha_i), \quad \alpha_i = \alpha_i' + 1.$

In accordance with the initial condition for the integral

$$\int_{-\infty}^{\infty} \exp(-\beta_i v_i^2 - \alpha_i)dv_i = 1$$

e.g. $\left(\frac{\pi}{\beta_i}\right)^{\frac{1}{2}} \exp(-\alpha_i) = 1$ and $\int_{-\infty}^{\infty} v_i^2 \exp(-\beta_i v_i^2 - \alpha_i)dv_i = \frac{kT}{m}$.

Hence $\beta_i = \frac{m}{2kT}$ and $\exp(-\alpha_i) = \left(\frac{m}{2\pi kT}\right)^{\frac{1}{2}}$.
Substitution into the expression for the probability leads finally to

$$p_i = \left(\frac{m}{2\pi kT}\right)^{\frac{1}{2}} \exp\left(-\frac{mv_i^2}{2kT}\right).$$

Solution 9.2. The potential of the chain depends on displacements of atoms $u_i = x_i - \langle x_i \rangle$ from their equilibrium positions $\langle x_i \rangle$:

$$U = \frac{1}{2}\sum_{i=1}(\alpha u_i^2 + \beta u_i u_{i-1}).$$

Perform a linear transformation of coordinates $u_i = S_{ik}u_k'$ to reduce this quadratic form to the diagonal shape: $U = \frac{1}{2}\sum_{i=1}\gamma_i u_i'^2$, where γ_i are the eigenvalues. Now, this potential corresponds to a system of independent oscillators with different vibration frequencies. It enables us to use a multiplicative form for the configuration of the integral of equation (9.14) with respect to new coordinates

$$\widehat{f}(\mathbf{u}') = \frac{1}{\widehat{Q}}e^{-\beta U(\mathbf{u}')} = \prod_{i=1}\sqrt{\frac{1}{\pi\beta\gamma_i}}\,e^{-\beta\gamma_i u_i'^2}.$$

By definition $\langle u_i'^2 \rangle = \frac{1}{2\beta\gamma_i}$ and $\left\langle u_i' u_j' \right\rangle = 0, \quad i \neq j$. Therefore

$$\langle u_i^2 \rangle = S_{in}S_{im}\langle u_n' u_m' \rangle = S_{in}S_{ni}^{-1}\langle u_n'^2 \rangle = \frac{1}{2\beta\gamma_i}$$

and

$$\left\langle u_i u_j \right\rangle = \frac{S_{in}S_{nj}^{-1}}{2\beta\gamma_n}, \quad i \neq j.$$

Thus we proved the conclusion obtained in the numerical simulations of chapter 4, on non-zero correlations between thermal displacements of atoms in a lattice. The temperature dependence $\left\langle u_i u_j \right\rangle \sim \sqrt{T}$ evidently follows from the theorem of the virial $U \sim K$.

Solution 10.1. The canonical distribution for the Hamiltonian of equation (10.5) is

$$f(\mathbf{x}\,,\,\mathbf{p},\,\gamma,\,p_\gamma)\exp\left[-\beta\left(\sum_{i=1}^{N}\frac{p_i^2}{2m_i} + \frac{p_\gamma^2}{2Q} + U + NkT\gamma\right)\right].$$

Hence the integration here over γ, p_γ leads to the canonical distribution with respect to the real phase space $\mathbf{x}$, $\mathbf{p}$ of equation (9.10).

Solution 11.1. Consider the expression

$$S(\alpha) = -k\sum_{i=1}^{N}(p_i \ \ln p_i - \alpha p_i + \beta p_i e_i),$$

where α and β are the Lagrange[1] multipliers. We have

$$\frac{\partial S}{\partial p_i} = -k(\ln p_i + 1 - \alpha) = 0.$$

Hence $p_i = \frac{1}{Q(E)} \exp(-\beta e_i)$ where $Q(E) = \sum_i \exp(-\beta e_i)$. Further

$$E = \sum_{i=1}^{N} p_i e_i = \frac{1}{Q(E)}\sum_i e_i \exp(-\beta e_i) = \frac{\partial \ln Q(E)}{\partial \beta}\bigg|_{e_1, e_2, e_3 \ldots}.$$

And finally

$$S_{\max} = k\sum_{i=1}^{N} p_i[\beta e_i + \ \ln Q(E)] = k\beta E + k \ln Q.$$

This value of the entropy corresponds to the canonical ensemble (compare with the result for the microcanonical ensemble $S_{\max} = k \ln N$). This result expresses the maximum entropy principle: a macrostate is described by a distribution over microstates where the values of the macroscopic observables are given by the expectation values and which also has maximum entropy, i.e., it contains no additional information about the system.

Solution 11.2. First calculate the density of states. The Hamiltonian of the system is written as $\mathscr{H}(\mathbf{x}, \mathbf{v}) = \frac{1}{2}\sum_{i=1}^{3}(v_i^2 + U(x_i - x_{i-1}))$.

For our purpose, it is sufficient to determine the density of states for $E < 0.5$ while $E/kT \gg 1$. Put $x_1 = k_i\delta, k_i = -K \div K, \ \ \delta = 1/K; x_2 = l_i\delta, l_i = 0 \div K, v_{1,2} = l_i\delta$. For $K > 10^3$ the accuracy of the results does not change. Of course, such a direct sampling is not effective; therefore we will use more speedy methods of free energy calculation below. Only two independent dynamical variables x_1 and x_2 take part in the dynamics, since $x_3 = 3 - (x_1 + x_2)$ and $v_3 = -(v_1 + v_2)$ due to momentum and center of gravity conservation. The density of states is shown in figure B.1. Note that the density of states in harmonic approximation enables analytical derivation. In particular, it is constant for one oscillator and drops down due to anharmonism of the pair potential.

Further, the calculation of the partition function of equation (9.13) yields the value of free energy of equation (11.1). Results are shown in figure B.2. In the limit

[1] As was mentioned, the Nosé–Hoover dynamics is non-Hamiltonian because it is a time transformation, so this proof is not rigorous. Without going into detail, the canonical distribution could be achieved by using a Nosé–Hoover chain of coupled thermostats.

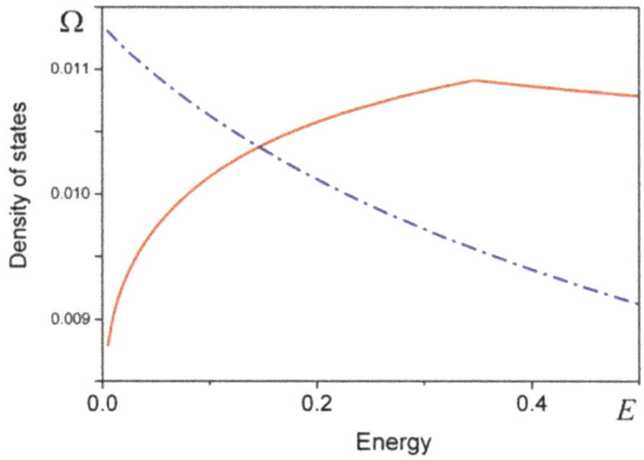

Figure B.1. Density of states (arbitrary units). The result for one oscillator is given by the dashed-dotted curve.

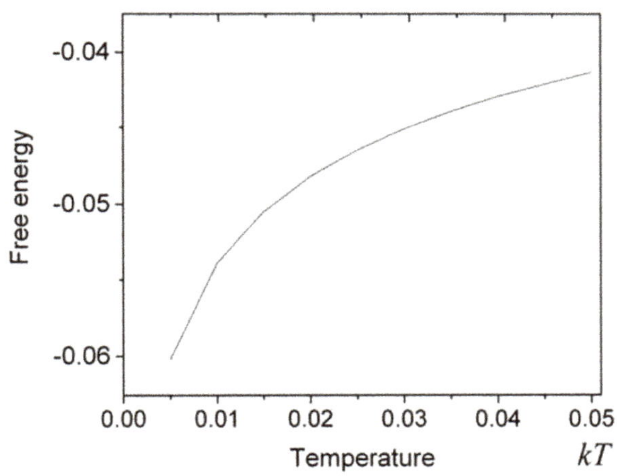

Figure B.2. Dependence of free energy on temperature.

$T \to 0$, the main contribution to the partition function gives the parabolic part of the potential where, as is known, $F \sim (1/\beta) \ln \beta$.

Solution 12.1. Since the entropy is additive $dS = \left(\frac{\partial S}{\partial E}\right)_{n_i} dE + \sum \left(\frac{\partial S}{\partial n_i}\right)_E dn_i$. A substitution here using the partial derivatives from (12.2) completes the proof.

Solution 12.2. The simulation program starts from a setting of coordinates according to uniform distribution inside the square $x_i, y_i \in -5 \div 5$ and velocities according to the normal distribution with the variance $\sigma^2 = 0.5kT$. Time evolution may be calculated with the help of the Verlet algorithm (see appendix A.1) for potentials equal to zero. The output of the number and the energy of particles at $t = 1000$ with the coordinates $x_i, y_i \in -2.5 \div 2.5$ gives the standard deviation near the mean values

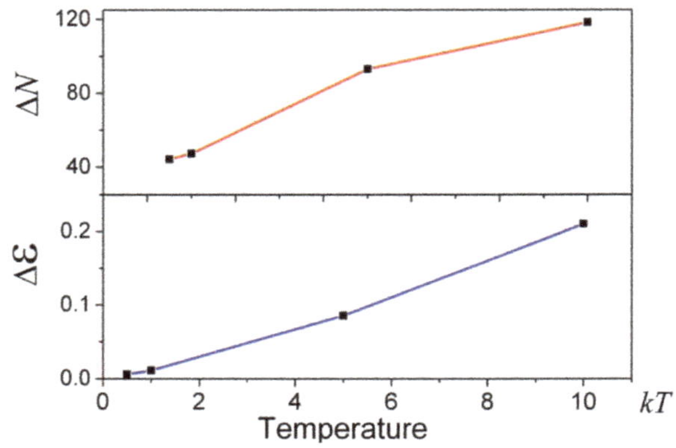

Figure B.3. Fluctuation of energy and the number of particles in a one fourth part of volume of the system of free particles.

(figure B.3). The analytical result establishes the dependence $\Delta N \sim kT$. Fluctuation of energy per particle $\Delta\varepsilon \sim kT/\sqrt{N}$ verifies independence of the statistical description of any part of a system.

Solution 13.1. The product PV is given in terms of the canonical partition function by

$$PV = \frac{V}{\beta}\frac{\partial \ln Q(V)}{\partial V} = \frac{V}{\beta Q}\frac{\partial Q}{\partial V}.$$

Averaging the product over an isothermal–isobaric ensemble yields

$$\langle PV\rangle = \frac{1}{\beta\widehat{Q}}\int_0^\infty dVe^{-\beta P^{\text{ext}}V}V\frac{\partial Q(V)}{\partial V}.$$

As was done for equation (9.19), the integration in parts gives

$$\begin{aligned}\langle PV\rangle &= \frac{1}{\beta\widehat{Q}}[e^{-\beta P^{\text{ext}}V}VQ(V)]_0^\infty - \frac{1}{\beta\widehat{Q}}\int_0^\infty dVQ(V)\frac{\partial Ve^{-\beta P^{\text{ext}}V}}{\partial V} \\ &= -\frac{1}{\beta\widehat{Q}}\int_0^\infty dVe^{-\beta P^{\text{ext}}V}Q(V) + \frac{P^{\text{ext}}}{\widehat{Q}}\int_0^\infty dVe^{-\beta P^{\text{ext}}V}VQ(V) \\ &= -kT + P^{\text{ext}}\langle V\rangle.\end{aligned}$$

Since $\langle PV\rangle$ and $P^{\text{ext}}\langle V\rangle$ are proportional to N, the extra kT term can be neglected in the thermodynamic limit and this equation becomes $\langle PV\rangle \approx P^{\text{ext}}\langle V\rangle$. This theorem, as well as other similar relations we encountered earlier, serve as a test to verify that a simulation procedure correctly samples a required ensemble.

Solution 14.1. Construct a set of equivalence $f = E, x = V, y' = \partial E/\partial V = -P$. The transformed function $E - (-P)V = E + PV = \Phi$ is the enthalpy.

Solution 14.2. If z is a function of x and y then the total differential is

$$dz = \left(\frac{\partial z}{\partial x}\right)_y dx + \left(\frac{\partial z}{\partial y}\right)_x dy.$$

If we move along a curve with $dz = 0$, where the curve is parameterized by x, then y can be written in terms of x, so on this curve

$$dy = \left(\frac{\partial y}{\partial x}\right)_z dx.$$

Therefore, the equation for $dz = 0$ becomes

$$0 = \left(\frac{\partial z}{\partial x}\right)_y dx + \left(\frac{\partial z}{\partial y}\right)_x \left(\frac{\partial y}{\partial x}\right)_z dx.$$

Since this must be true for all dx, rearranging terms gives

$$\left(\frac{\partial z}{\partial x}\right)_y = -\left(\frac{\partial z}{\partial y}\right)_x \left(\frac{\partial y}{\partial x}\right)_z.$$

Dividing by the derivatives on the right-hand side gives the cycle rule

$$\left(\frac{\partial x}{\partial y}\right)_z \left(\frac{\partial y}{\partial z}\right)_x \left(\frac{\partial z}{\partial x}\right)_y = -1.$$

Solution 15.1. It is seen from equation (16.11) that the thermal capacity $C_V = (\partial E/\partial T)_V$ coincides with that for an ideal gas. Hence van der Waal's equation gives

$$C_p - C_V = \frac{N}{1 - \dfrac{2Na}{TV^3}(V - Nb)^2}. \tag{B.1}$$

Solution 16.1. Take the exterior derivative of the equation (11.12): $dE = T\,dS - P\,dV$. Since $d(dE) = 0$, we obtain $0 = dT\,dS - dP\,dV$. This leads to the determinant $\frac{\partial(T,S)}{\partial(P,V)}$. The Maxwell relation now follows directly $\left(\frac{\partial S}{\partial V}\right)_T = \frac{\partial(T,S)}{\partial(P,V)} = \frac{\partial(P,V)}{\partial(T,V)} = \left(\frac{\partial P}{\partial T}\right)_V$.

Solution 17.1.

$$\mu_{JT} = \frac{(2a/kT)(1 - b\rho)^2 - b}{c_P[1 - (2\rho/kT)(1 - b\rho)^2]} - a\rho^2$$

Solution 21.1. For two phases in equilibrium $G_1 = G_s$ and $dG_1 = dG_s$ for infinitesimal change in T and P (so that the system remains in equilibrium). From equation (14.4) follow $dG_l = V_l dP - S_l dT$ and $dG_s = V_s dP - S_s dT$ or $V_l dP - S_l dT = V_s dP - S_s dT$ and further $\left(\frac{dP}{dT}\right)_{\text{eq}} = \frac{S_{\text{eq}} - S_l}{V_s - V_l} = \frac{\Delta S}{\Delta V}$. At equilibrium $\Delta G = \Delta\Phi - T\Delta S = 0$, so $\Delta S = \Delta\Phi / T$. Substitution into the derivative yields the *Clapeyron equation.* It says that when pressure changes at a fixed temperature their free energies should be increased by different amounts. The only way to maintain equilibrium at different pressures is to change the temperature as well.

Solution 22.1. The energy of the system is equal to $E \doteq -(N_+ - N_-)\mathcal{H}$. Calculation analogous to equation (22.9) yields

$\ln \Omega(E) = N \ln 2N - \frac{1}{2}(N - E')\ln(N - E') - \frac{1}{2}(N + E')\ln(N + E')$, where $E' = E/\mathcal{H}$. From this $E = -N\mathcal{H}\ \text{th}(\beta\mathcal{H})$.

Lightning Source UK Ltd.
Milton Keynes UK
UKHW05n1615231018
331042UK00004BB/142/P